Ten-Minute Real World Math

by Scott McMorrow
illustrated by Phillip Chalk

> This book is dedicated to
> Peter McGowan

Publisher: Roberta Suid
Editor: Hawkeye McMorrow
Design & Production: Scott McMorrow

Other Monday Morning publications in this series:
*Ten-Minute Grammar Grabbers, Ten-Minute Editing Skill Builders,
Ten-Minute Thinking Tie-ins, Ten-Minute Real World Reading,
Ten-Minute Real World Science, Ten-Minute Real World Writing*

Entire contents copyright © 1997 by Monday Morning Books, Inc.
Box 1680, Palo Alto, CA 94302

For a complete catalog, please write to the address above,
or visit our Web site: www.mondaymorningbooks.com
e-mail: MMBooks@aol.com

monday morning®

Monday Morning is a registered trademark of
Monday Morning Books, Inc.

Permission is hereby granted to reproduce
student materials in this book for non-commercial
individual or classroom use.

ISBN 1-57612-021-X

Printed in the United States of America
987654321

Contents

Introduction 4

Money Math
In the Budget 6
Budgeting Basics 7
Checking Account Math 8
Check Patterns 9
Recycling Pays 12
Recycling Reasoning 13
Stock Market Math 14
Ups and Downs 15
Tipping Etiquette 16
Restaurant Math 17
Recipe Revision 18
Cooking with Math 19
25 Cents A Day 20
Savings Account Math 21
Taxing Multiplication 22
Income Tax Returns 23
Shopping Smart 24
Shopping with Coupons 25
Buying on Credit 26

Autobiographical Math
Walking Speed Calculating 27
Equator Math 29
Breath Count 30
Growth Charting 31
Favorite Foods Graphing 32
Bony Bar Graphing 33
Heartbeat Multiplication 34
Heart Rate Worksheet 35
Personal Percents 36
Personal Percents Pie Charting 37
Hair Count Estimating 38
Eye Blinks Estimating 39
Pace Counting Conversion 40
Leg-Up Geometry 41
Sun Protection Factors 42
School Days 43
Time Budget 44
Time Log 45
Family Percents 46
Family Percents Graphing 47

Mental Math
Rate it Right 48
Simple Surveying 49
Kid Count Estimating 50
How Many Books? 51
Counting Haiku 52
Map Reading Math 53
How Far? 54
Drawing With Shapes 55
Drawing to Scale 56
Estimating Distance 57
Divide by Eye 58
Ballpark Geometry 59
Measuring with Paper 60
Tool-of-Thumb Scaling 61
Fingertip Measuring 62
Fingertip Measuring Chart 63
Syllable Counting 64
Calculating Readability 65

Numbers in the News
24-Hour Time 66
It's a Watch... It's a Compass 67
Greenwich Mean Time (GMT) 68
Weather Map Data 69
Census Report 70
Temperature Tracking 71
Mars Mission Math 72
More Martian Math 73
Global Positioning System 74
Shape Locators 75
Metric Conversions 76
Computer Memory Math 77
Sports Statistics 78
Team Tracking 79
Euro Dollars 80
Currency Exchange 81
Paying for Television 82

Resources
Mathematician's Toolbox
    Mathematician's Glossary 83
    Conversion Tables 87
    General Formulas 89
    Graph Pattern 91
    Pie Chart Pattern 92
Web Extensions 93
Math Skills Index 95

INTRODUCTION

Mathematics surrounds us in our everyday lives: at the checkout counter, at the bank, and in our cars. One of the often overlooked strengths of math is how it can be used to develop cultural literacy (such as knowing what Greenwich Mean Time is, or understanding the structure of haiku). While it's true that math is everywhere, it is even truer that a firm grasp of analytical concepts will broaden our view of the world and the people in it.

Secrets to Math Motivation

The perennial question asked by students ("Why do I need to know math?"), is best answered through real world examples of topics that are informative and exciting. Getting students interested in math involves allowing them to explore math at a level they can relate to.

The Ten-Minute Philosophy

A student's comfort level with math is directly related to the amount of time spent practicing math and the variety of problems encountered. The exercises in this book were created to maximize the quality of practice time while offering a wide range of interesting subjects. Students will develop and strengthen the skills needed to successfully face the challenges of more advanced math topics. These skills include:

- logical thinking
- analyzing data
- graphing
- map reading
- estimating
- measuring
- problem solving
- geometry
- addition/subtraction
- multiplication/division
- ratio/percents
- perimeter/area

The Lessons

Each lesson starts with background information that introduces the topic and defines the use of the subject. Step-by-step instructions explain each lesson's process. This format allows students to quickly grasp the concept being taught while giving them the opportunity to hone their problem-solving skills.

Beyond Ten Minutes

An extension activity follows each ten-minute lesson. These longer activities enable students to reinforce the skills developed in the shorter lesson.

The extensions will enhance the students' cultural knowledge through the use of analytical skills. Whether done as in-class assignments or homework projects, the extensions are powerful tools that aid students in acquiring a well-rounded education.

Bonus Section

The Resource section contains a content-rich Mathematician's Toolbox. This reproducible student guide contains the following essential tools:

- Mathematician's Glossary: a listing of terms and jargon encountered throughout the various math disciplines (arithmetic, geometry, trigonometry, and more).
- Conversion Tables: Metric system equivalents, volume to weight conversions, Celsius and Fahrenheit formulas, and more.
- General Formulas: standard geometric shapes, areas, and more.

The **Web Extensions** segment of the Resource section contains valuable World Wide Web addresses (URLs) that connect the lessons and activities to the Internet. These URLs provide excellent reinforcement of the lessons in this book, and can be used as starting points for Internet projects.

A **Math Skills Index** identifies which areas of math (subtraction, addition, graphing, multiplication, etc.) are used in the lessons.

Where to Begin

The lessons in *Ten-Minute Real World Math* are designed to be used independently, except where noted. Focus on activities that complement the math skills being taught in your classroom, or explore specific topics of interest to your students.

IN THE BUDGET

Budgets are useful tools for determining how money is spent. A total sum of money is divided into smaller amounts that will be used for specific needs.

MATERIALS:
Paper, pencil

DIRECTIONS:
1. While explaining the concept of a budget to the class, be sure to tell the students that families, schools, cities, states, and governments use budgets to plan spending.
2. Write the following on the board:

	Annual Police Department Budget		
Total Money to spend	% of total for salaries	% of total for training	% of total for cars
750,000	50%	30%	20%

3. Have the students calculate the amount of money this police department will spend on salaries (750,00 x .5 = 375,00 dollars), training (225,000 dollars), and cars (150,000 dollars).

EXTENSION:
Strengthen students' budgetary understanding by having them complete the Budgeting Basics activity on the following page. Students can present their completed budgets to the class.

Budgeting Basics

Imagine you have a job working as a bicycle messenger, and you get paid 10 dollars an hour.

1. How much do you get paid for working 40 hours in one week?

2. Using the answer from question 1, how much do you earn in one year if you work 52 weeks in that year?

3. Using the annual income calculated in question 2, determine how much you can spend for housing, food, clothes, savings, and fun. The following are the percentages you will spend in each area:

Housing 20%   Food 30%   Clothes 10%
Savings 25%   Fun 15%

CHECKING ACCOUNT MATH

Setting up a checking account, keeping the account balanced, and monitoring check writing are vital skills in today's world.

DIRECTIONS:
1. Explain to the students that a checking account is a type of bank account that allows someone to make purchases by writing checks instead of using money. This system helps people keep track of their spending.
2. Also, explain that checks draw from money existing in a person's bank account, and that each person is responsible for keeping track of all checks written for her or his account.
3. On the board, draw the page from the check register shown below.

Date	Check #	Transaction	Amount	Deposit	Balance
05/17		paycheck		1263.54	1263.54
05/21	100	Phone Bill	32.00		
05/23	101	Electric Bill	65.00		
05/30	102	Gas Bill	23.15		

4. Have students copy the register onto their papers, and perform the addition and subtraction needed to balance the check register. You might demonstrate the work on the board to help them get started.

EXTENSION:
Using the patterns on the following pages, create a checking system for the students. Start each student with the same balance. During the course of a week, issue "bills" to the students for various things, such as classroom electricity, building maintenance, and school supplies. Have the students write checks for these bills, and keep the check register balanced. At the end of the week, collect the registers and review the students' work.

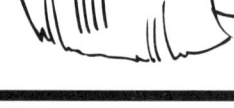

Check Patterns

Center County Bank Check # _____

Date: _____

PAY TO THE ORDER OF _____ $ _____

_____ DOLLARS

MEMO _____   _____

Center County Bank Check # _____

Date: _____

PAY TO THE ORDER OF _____ $ _____

_____ DOLLARS

MEMO _____   _____

Center County Bank Check # _____

Date: _____

PAY TO THE ORDER OF _____ $ _____

_____ DOLLARS

MEMO _____   _____

Register Patterns

Check Register				
Date	Check #	Transaction	Amount	Balance

Check Register				
Date	Check #	Transaction	Amount	Balance

Bill Patterns

Billing Summary		
Date	Description	Amount Due

Please Make Check Payable To:
_____

Account Number: 4363-9-8905

Billing Summary		
Date	Description	Amount Due

Please Make Check Payable To:
_____

Account Number: 4363-9-8905

RECYCLING PAYS

The need for recycling grows as resources become limited and landfill disposal space gets harder to find. Understanding the value of recycling is the first step in the fight to reduce unnecessary waste.

MATERIALS:
Paper, pencil

DIRECTIONS:
1. Describe to the class how paper recycling works. People save all the paper that they otherwise would throw away. This paper is gathered at recycling centers and it is chopped up, bleached, and made into blank paper again.
2. If an average of 72 million tons of paper is used in one year, and 26% of this is recycled, ask the students how many tons of paper are recycled? Write the following on the board to help:

> Amount recycled/72,000,000 = 26%,
> Amount Recycled= (.26) x 72,000,000

(Answer = 18,720,000 tons)

EXTENSION:
Expand students' knowledge of recycling with the Recycling Reasoning graphing activity on the following page.

Recycling Reasoning

Use the following information to calculate the quantities of materials that are recycled.

Material	Amount Used	Percent Recycled
Glass	13,000,000 tons	6%
Metal	12,000,000 tons	3%
Plastic	14,000,000 tons	1%

For example:
Amount of Plastic recycled = 1% x 14,000,000 = (.01) x 14,000,000 = 140,000 tons

Use your answers to complete the bar graph below.

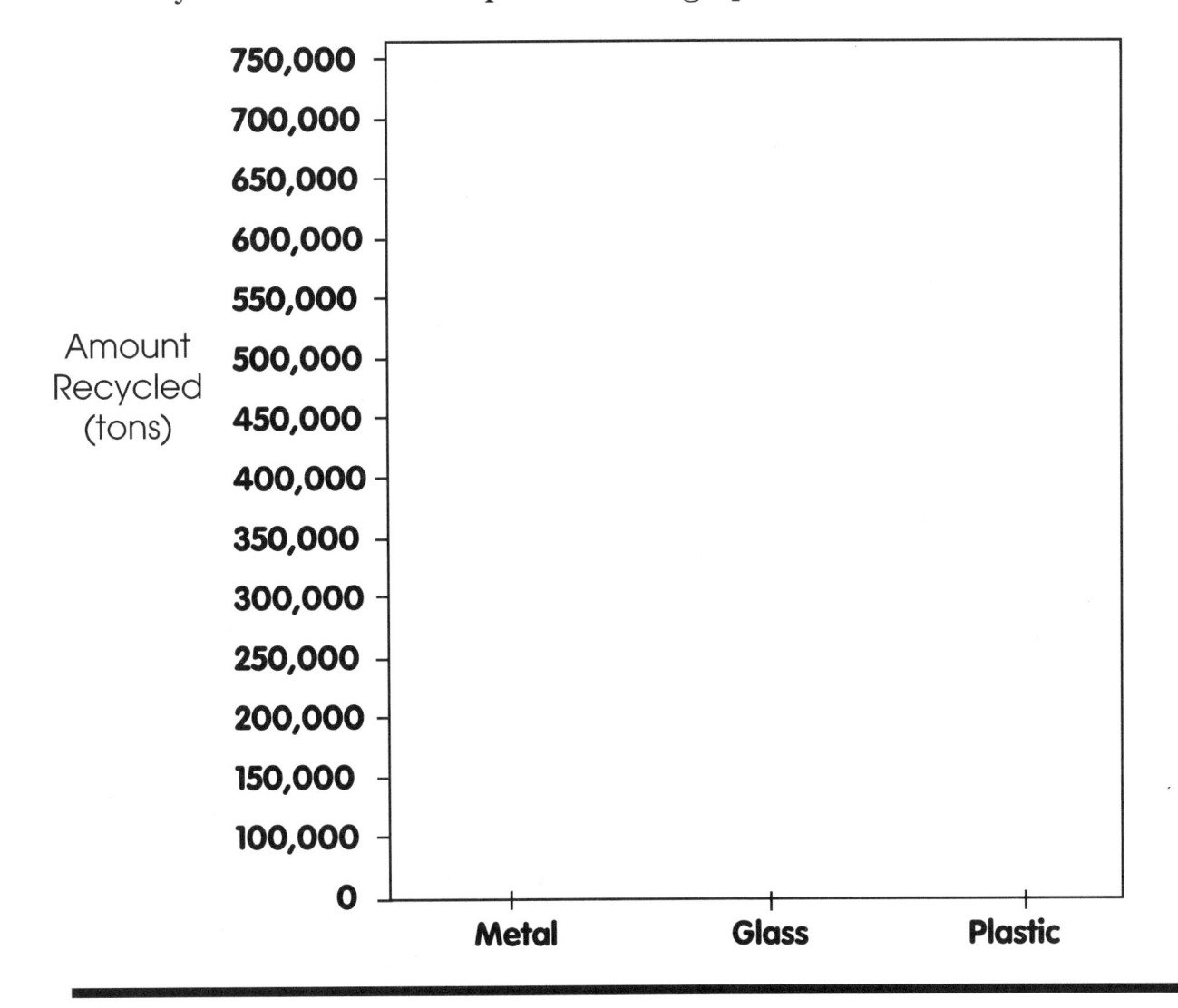

STOCK MARKET MATH

Stock markets are places where people speculate how well, or poorly, a company is going to perform. Understanding the basics of how these markets work is essential for successful investments.

DIRECTIONS:
1. Explain the concept of a stock market to the class.
People buy shares (pieces of ownership) of a company. When the value of these shares goes up, people sell them for more than they paid. If the value of the shares goes down, the investor can lose money.
2. Write the following on the board:

Date	52-week high low	Stock	Closing Price
01/30/97	57.25 50.05	MMBooks	56.80

(This represents the information for a single share of stock)

3. Have the class calculate how much 500 shares of MMbooks would be worth at closing on 01/30/97. (Answer = 28,400 dollars)

EXTENSION:
Strengthen students' understanding of the stock market by having them complete the Ups and Downs activity on the following page. Have the students present their results in class.

Ups and Downs

Daily newspapers list stock market information in their business section. Companies are usually listed by their initials. For example, McDonald's is listed as McDnld. A "How to Read the Stock Data" guide can be found on the stocks page.

Using a daily newspaper, select a stock and chart its closing price for five days. Use the table below to keep track of the stock's performance.

DATE	STOCK	CLOSING PRICE

NEWSPAPER _____

TIPPING ETIQUETTE

In many countries, it is considered proper etiquette to add extra money to the bill for a meal. This additional payment, or tip, is compensation for good service.

DIRECTIONS:
1. Explain to the students that calculating a tip is based on a percentage of the total cost of the meal. Tips usually range from 10% to 20% of the cost.
2. On the board, draw the restaurant check shown below.

2	slice of cheese pizza	1.25
		1.25
2	lemonade	1.05
		1.05
total		4.60

3. Have students figure out a 20% tip. Explain that a simple way to calculate 20% is to first figure 10% of the bill by moving the decimal one place to the left (.46), and then multiply this amount by two (total tip = .92).
4. Ask the students what the total cost of the bill is, including tip (Answer = 5.52 dollars).

Optional:
1. Have the students figure out a 15% tip. Explain that a simple way to calculate 15% is to use the 10% calculation (.46), then add half of this amount (.23) to the 10% (total tip = .69).
2. The students can now compute the total cost of this bill, including tip (Answer = 5.29 dollars).

EXTENSION:
In class, or for homework, duplicate the following page and have students complete the Restaurant Math activity.

Restaurant Math

Name:_____

Date:_____

Order for two people, including: drinks, bagels, and fruit. On a separate piece of paper, calculate the total cost of the meal. Then figure out a 15% tip and a 20% tip.

Pete's Bagel & Juice Shop
Menu

Bagels
Bagel	1.00
Bagel with Cream Cheese	1.75

Drinks
Orange Juice	1.50
Milk	1.00

Fruit
Banana	.75
Apple	.60

Okay, 15% of $21.50 is....

RECIPE REVISION

Cooking recipes usually make a certain quantity of food. Cooks often adjust the ingredient amounts to match the number of people they are cooking for.

MATERIALS:
Paper, pencil

DIRECTIONS:
1. Tell the class that many meals are prepared by following recipes. Sometimes the amounts of ingredients are changed if the recipe doesn't make the exact quantity of food needed.
2. Write the following recipe on the board:

<u>10 Pancakes</u>
1 1/2 cups flour (340 grams)
2 1/2 teaspoons baking powder (10 grams)
3/4 teaspoon salt (5 grams)
1 egg
1 1/4 cups milk (.5 liters)
3 tablespoons salad oil (1 liter)

One method of solving this activity is to multiply each ingredient amount by the number of students in the class. Then divide each ingredient amount by 10.

3. Have the students silently count the number of people in the classroom.
4. Assume each person can eat one pancake. Have the students adjust the ingredient amounts to make enough pancakes for everyone.

EXTENSION:
Challenge the students' ability to convert by having them calculate a pancake recipe for:
a) their family, b) one person, and
c) 300 people.

COOKING WITH MATH

Oven temperature is given in degrees Fahrenheit, or degrees Celsius, depending on what country you live in. The ability to convert temperatures from one degree type to the other can be useful in cooking.

MATERIALS:
Paper, pencil

DIRECTIONS:
1. Tell the class that temperatures can be expressed in two different units of measure, degrees Celsius and degrees Fahrenheit.
2. Write the following conversion formulas on the board:

> Celsius to Fahrenheit (9/5 x degrees Celsius) + 32 = degrees Fahrenheit
> Fahrenheit to Celsius (degrees Fahrenheit - 32) x 5/9 = degrees Celsius

3. Have the students convert an oven temperature of 425 degrees Fahrenheit into degrees Celsius.
4. Have the class convert an oven temperature of 150 degrees Celsius into degrees Fahrenheit.

EXTENSION:
Have the students research the freezing point and boiling point temperatures of water. Students should report their findings in degrees Fahrenheit and degrees Celsius.

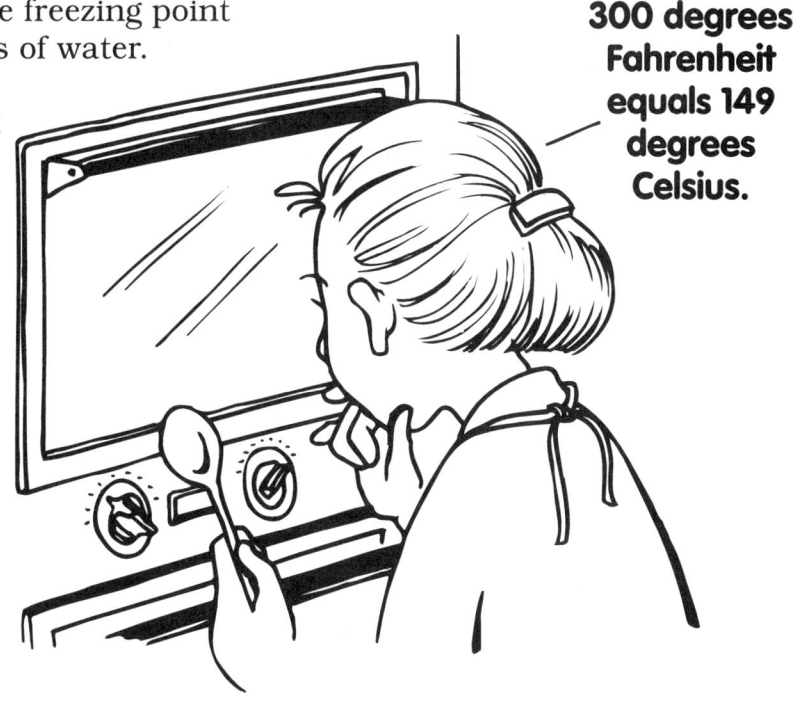

300 degrees Fahrenheit equals 149 degrees Celsius.

25 CENTS A DAY

Learning the value of saving money will benefit students for their entire lives. Saving even a small amount of money on a regular basis can add up.

MATERIALS:
Pencil, paper

DIRECTIONS:
1. Explain to the class that a bank savings account is a place to keep money. Adding money to the account is called making a deposit.
2. Tell the students they are going to calculate the amount of money they would have by saving 25 cents a day for a year.
3. Ask the class how many days there are in one year? (365)
4. Have the class multiply 365 by .25 (91.25 dollars). This would be their savings for one year.

EXTENSION:
Strengthen students' appreciation for savings accounts by having them calculate: a) saving 50 cents a day, b) saving 1 dollar a week, and c) saving 5 dollars a month. All calculations should be based on saving for one year.

SAVINGS ACCOUNT MATH

Most banks give interest to customers who put money in savings accounts. Interest rates vary, but an average is 4% annually. Students need to be aware of interest bearing accounts so they can maximize their savings.

MATERIALS:
Paper, pencil

DIRECTIONS:
1. Explain to the class that banks pay customers by offering interest on savings accounts. This payment is for the bank's use of their money. The amount of money paid is based on a percentage of the account's savings.
2. If the bank pays interest of 4% annually, have the students calculate the percent of interest paid monthly.
(Answer = 4%/12 months = .33%)
3. Have the class imagine they start with 100 dollars in a savings account. Each month the bank calculates .33% of the amount saved and adds this money to the account.
4. Have the class calculate how much interest the bank will give after one month. (Answer = 100 x .0033 = .33)
5. The following month the bank gives another .33% of the accounts money. Have the class calculate the amount of money in the account after two months. (100.33 x .0033 = .33, Answer = 100.33 + .33 = 100.66 dollars)

EXTENSION:
Develop students' understanding of interest bearing accounts by having them calculate 12 months of interest for the account above. (104 dollars)

TAXING MULTIPLICATION

Taxes are used to raise revenue for products and services used by citizens. Federal, state, and local taxes are placed on a variety of items, such as income, property, and retail products.

MATERIALS:
Paper, pencil

DIRECTIONS:
1. Ask the class if they know what taxes are. After hearing their responses, be sure they understand that taxes are imposed by Federal, state, and local governments. The money gathered through taxes pays for such things as building and maintaining highways, operating police and fire departments, and funding public libraries.
2. Write the following on the board:

money spent on new library books per year	number of taxpayers in town
100,000 dollars	20,000

3. Have the students calculate the amount of money a taxpayer pays for new library books in one year.
(100,000/20,000 tax payers = 5 dollars/tax payer)
4. Ask the class to brainstorm ideas on what types of taxes would be used to pay for library books. (Some examples might be: property taxes on houses, state and local income taxes, and sales tax.)

EXTENSION:
Help the students develop their understanding of taxes by having them research the sales tax for their area. They can visit or call a local merchant and ask what the tax rate is, what types of goods are taxed, and what types of products are not taxed. Students can present their findings to the class.

INCOME TAX RETURNS

Income taxes are collected by different governments and are based on the amount of money a person earns in one year.

MATERIALS:
Paper, pencil

DIRECTIONS:
1. Explain to the class that workers pay a portion of the money they earn to the government. This payment is based on the amount of money, or income, a person earns in one year.
2. Tell the students that there are two levels of tax payments made. One payment goes to the Federal government, and the other goes to their state government.
3. Have the class calculate a person's annual income based on the following information:

dollars earned per hour	hours worked per week	weeks worked per year
10	40	52

(Answer = 10/hr x 40 hr/wk x 52 wks/yr = 20,800 dollars/yr)

4. Tell the class that single people pay approximately 33% of their income for Federal and state taxes. Have the class determine the amount of money a single person has left after paying taxes. (20,800 x .33 = 6,864. 20,800 - 6,864 = 13,936 dollars left after taxes)

EXTENSION:
Duplicate enough copies of a tax form for each student (these are available at libraries and the Post Office). Go over the form in class, explaining each line to the class. Have the class fill out the form using the income from above and the 33% tax approximation.

SHOPPING SMART

Comparing prices when food shopping is a great way to save money. One thing to compare is the price per pound (kilogram) of a product.

MATERIALS:
Paper, pencil

DIRECTIONS:
1. Explain to the class that many foods found in stores have the cost per pound (kilogram) listed. This information is useful when comparing the prices of similar products.
2. Have the students imagine they are comparing two types of their favorite cereals. The following information is listed:

	Cereal 1	Cereal 2
Total Cost	3 dollars	4 dollars
Price per pound	2 dollars/lb	1 dollar/lb

3. Ask the class to pick the cereal that is the better buy.
(Answer: Cereal 2. Although the cost per box of Cereal 2 is higher, it is less expensive by weight. Cereal 1 has a higher cost per pound, so Cereal 1 must have less cereal in the box to have the low price per box.)

EXTENSION:
Help students develop smart shopping skills. Duplicate the Shopping with Coupons activity for use as a homework assignment. If the students have difficulty finding coupons, ask a local store to donate unsold Sunday newspapers to the class. Stores can also donate coupon books.

Shopping with Coupons

Coupon shopping is a way to save money. Newspapers usually have a coupon section in their Sunday edition. Also, supermarkets sometimes have coupon books at the store. Using the form below, look through the newspaper's coupon section and calculate the amount of money saved with coupons.

Find as many of the items as possible. If an item is listed on the form, but you can't find a coupon for it, cross out the item and write in one that you have a coupon for.

Item	Coupon Value
canned vegetables	
soap	
ice cream	
cereal	
Total Savings	

BUYING ON CREDIT

Lending rates are surcharges attached to loans and credit advances from the lending institution. It's smart to do a little research before getting a credit card or bank loan. Shopping around for a low interest rate can save money.

MATERIALS:
Paper, pencil

DIRECTIONS:
1. Ask the students if they know what a credit card is. Explain that a credit card is used to buy something now and pay for it later. The interest rate on the credit card is the amount of extra money that has to be paid when buying on credit.
2. Have the students calculate the amount of interest they would pay, after one year, for the following: (Answer = 9 dollars)

purchase	annual interest rate
60 dollar shoes	15%

If we buy it on credit, we pay $5 a month for seven months

3. Have the class determine the total paid for the shoes if they paid at the end of one year..
(60 + 9 = 69 dollars)

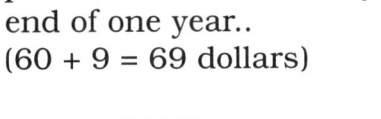

EXTENSION:
Have the students calculate the payments for the following, assuming monthly payments of 30 dollars each:

purchase	monthly interest rate
60 dollar shoes	1.25%

(Answer: <u>1st month</u>, .0125 x 60 = .75 = interest owed, 60 + .75 = 60.75 = total with interest, 60.75 - 30 = 30.75 = amount owed after first month. <u>2nd month</u>, .0125 x 30.75 = .38 = interest owed, 30.75 + .38 = 31.13 = total with interest, 31.13 - 30 = 1.13 owed after second month. <u>3rd month</u>, pay remaining 1.13.

WALKING SPEED CALCULATING

Speed, also known as velocity, is calculated by knowing how much time it takes to travel a given distance. Velocity is usually expressed as distance over time, such as miles per hour (mph), or kilometers per hour (kph).

MATERIALS:
Tape, measuring tape, watch with second hand, paper, pencil

DIRECTIONS:
1. In preparation for this activity, measure 20 feet (7 meters) on the floor. This can be done in the classroom, hallway, or playground. Use the tape to mark each end of the 20 feet (7 meters).
2. Have students place their toes on the edge of the beginning tape mark.
3. While standing at the end tape mark, tell the students to start walking toward you. Note the time you told the class to "go."
4. As each student crosses the end mark, count off the number of seconds that elapsed. Students should write this time down on their papers.
5. Have the students divide the number of seconds into the 20 feet (7 meters). This is their walking velocity in feet (meters) per second.
6. Have the class multiply feet per second by 0.68 to convert to mph (meters per second by 3.6 to convert to kph).

EXTENSION:
The formula for velocity (V = distance/time) can be rewritten to calculate how much time is needed to go a known distance (time = distance/velocity). Have the students calculate how much time they would need to: a) walk from their homes to school, b) "walk" from Earth to the moon, and c) walk from the base of Mt. Everest to the summit.

Couldn't you slow down a little?

Equator Equivalents

The equator is an imaginary circle around the earth located halfway between the North and South Poles. Measuring 24,000 miles (38,600 kilometers), this ring divides the earth into two halves: the Northern Hemisphere and the Southern Hemisphere.

DIRECTIONS:
1. Explain to the class that the distance between a city and the equator can be calculated from the number of degrees the city is from the equator. The equator is at zero degrees. Latitude lines are given in the number of degrees they are from the equator.
2. Ask the students why it might be useful to know that each degree of latitude equals 69 miles (111 kilometers)? You can use this fact to determine a location's distance from the equator.
3. Write the following on the board:

- Dunbar, Scotland: latitude 56 degrees north
- 1 degree of latitude equals 69 miles (111 km)

(note: degrees are given north or south from the equator.)

4. Have the students calculate Dunbar's distance from the equator. (Answer = 56 x 69 miles (111 kilometers) = 3,864 miles (6,216 km)

EXTENSION:
Have the students reinforce the concepts of latitudes, and distances from the equator, by completing the Equator Math activity on the following page. Duplicate one copy of the page for each student.

Medicine Hat, Canada is fifty degrees north of here!

Equator Math

Using the information below, calculate the city's distances from the equator. Determine these distances in both miles and kilometers.

New Orleans, United States	30 degrees north
Dunbar, Scotland	56 degrees north
Medicine Hat, Canada	50 degrees north
Wagga Wagga, Australia	35 degrees south

- 1 degree of latitude equals 69 miles (111 kilometers)

Place	Latitude	Miles from Equator	Kilometers from Equator

For further equator fun, research the latitude of your home town and another country's capital. Calculate the distances these are from the equator. Libraries are a good place to look for maps that have degrees of latitude. If the exact latitude is not shown on the map, find the nearest latitude shown.

BREATH COUNT

Most of us take breathing for granted because drawing breath is one of the automatic reflexes of the human body.

MATERIALS:
Watch with second hand, pencil paper

DIRECTIONS:
1. Tell the class they are going to calculate the number of breaths they take in one minute.
2. With your timer in hand, tell the students to start counting the number of breaths they take. Students should breathe at their normal rate.
3. After 15 seconds have elapsed tell the class to stop counting. Students should write the number of breaths on their papers.
4. Have the class multiply this number by 4. This is the number of breaths they take in one minute.

EXTENSION:
For homework, have the students calculate: a) the number of breaths they take in one year, b) the number of breaths they have taken in their lives, and c) the number of breaths William Shakespeare took.

How old were you, Mr. Shakespeare?

I lived for 52 wonderfully creative years.

GROWTH CHARTING

Students' awareness of themselves can be developed through charting the growth of their bodies.

MATERIALS:
Tape measure, paper, pencil

DIRECTIONS:
1. At the beginning of the school year, measure the height of each student. In preparation for this, use a tape measure to mark out a ruler in a convenient place, such as the doorway, a wall, or a long strip of paper.
2. Have each student stand next to the ruler while another student reads off the height. A third student should write down the height measurement in a notebook. When all the students are measured, store the notebook in a safe place.
3. At the end of the school year, repeat the measuring process.
4. Have the students calculate the change, if any, in their heights.

EXTENSION:
Have the students work on a short term Growth Charting activity during the school year. Charts could include hair growth or fingernail growth. Be sure the class keeps an accurate log of the results.

You've grown since last time.

FAVORITE FOODS GRAPHING

Bar graphs are useful for analyzing data, demonstrating relationships between varying information, and illustrating numerical concepts in a visual format.

DIRECTIONS:
1. Explain to the students that a bar graph is an effective way to present data in an easy-to-understand format.
2. On the board, draw the grid-layout shown below.

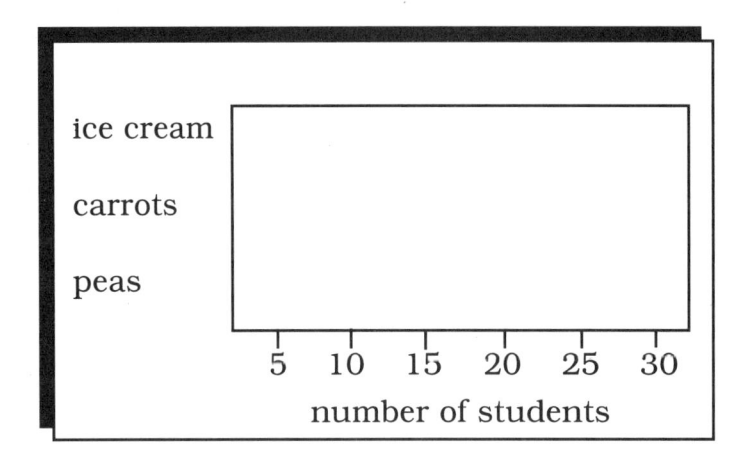

ice cream

carrots

peas

5 10 15 20 25 30
number of students

3. Ask the students the following questions and chart their answers:
   - "How many students like ice cream?"
   - "How many students like carrots?"
   - "How many students like peas?"
4. Have the students interpret the bar graph by asking questions about the relationships between the three data, such as:
   - "How many more students like ice cream than peas?"
   - "How may answers were given for all three questions?"
   - "What was the least favorite type of food?"

EXTENSION:
Have the students explore bar graphing by doing the activity on the next page.

Ten-Minute Real World Math ©1997 Monday Morning Books, Inc.

Bony Bar Graphing

Use the bar graph below to answer the questions on a separate piece of paper. Do your work in pencil so that you can easily make corrections.

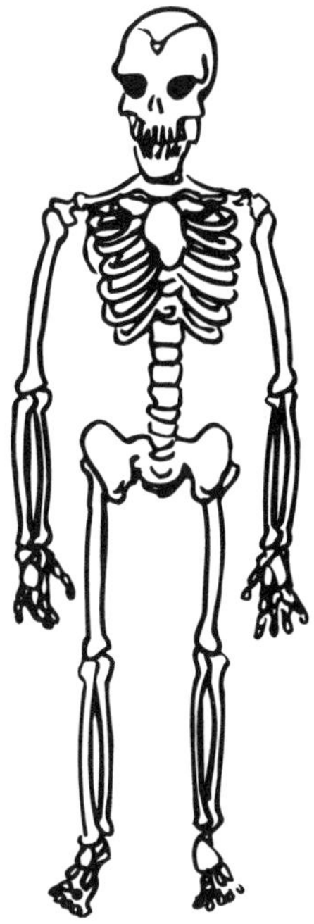

Bone Groups

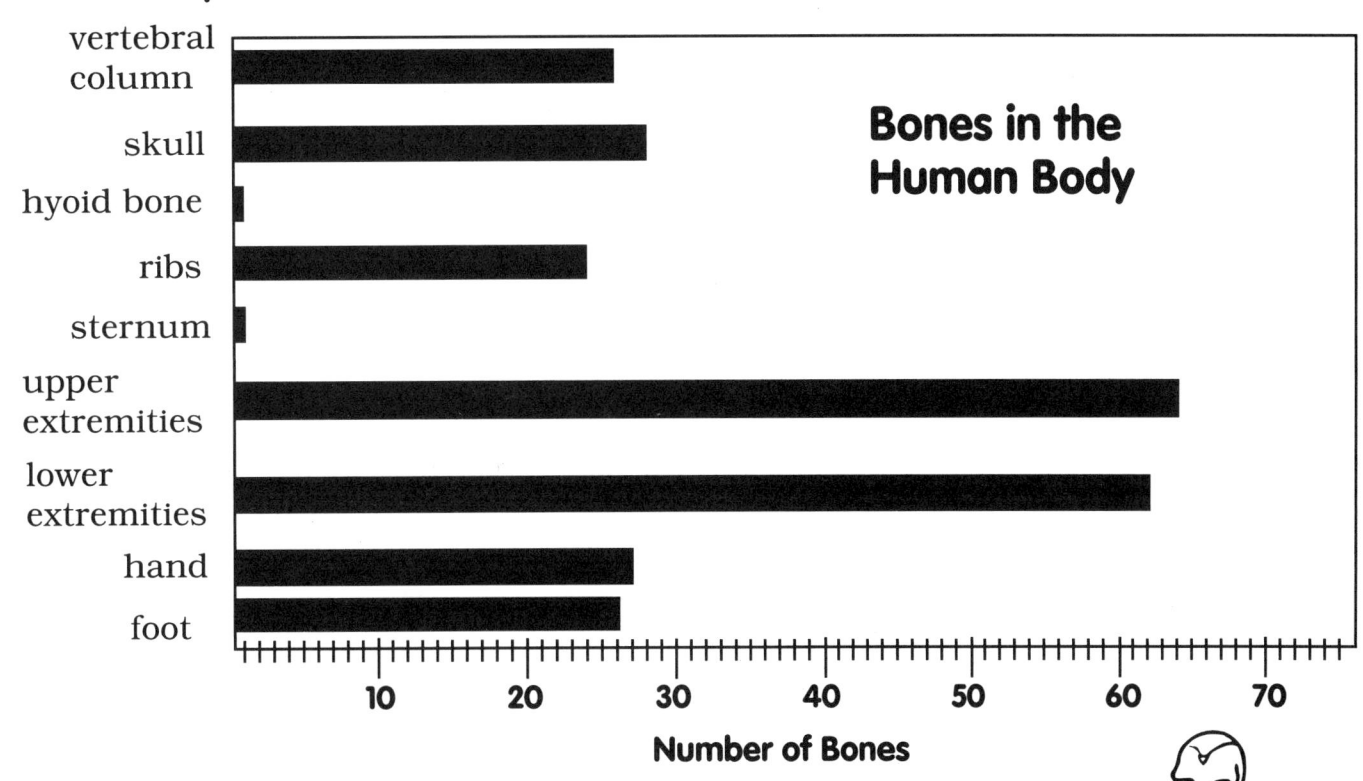

Bones in the Human Body

(Bar graph showing approximate values:)
- vertebral column: ~26
- skull: ~28
- hyoid bone: ~1
- ribs: ~24
- sternum: ~1
- upper extremities: ~64
- lower extremities: ~62
- hand: ~27
- foot: ~26

Number of Bones

1. How many more bones are in the upper extremities than in the vertebral column?
2. What's the total number of bones in both the upper and lower extremities?
3. What is the total number of bones in the ribs, sternum, and hyoid groups?
4. How many bones are there in the skull?
5. How many bones are there in a human skeleton?

Heartbeat Multiplication

The number of heartbeats in one minute is generally called the heart rate. Measuring heart rate gives useful information about how the heart is working.

MATERIALS:
Watch with a second hand, pencil, paper

DIRECTIONS:
1. Explain to the students that they can take their pulse in several places. One pulse point is located on the inside of the wrist on the thumb side. Another pulse point can be found by gently pressing under the jaw bone, just to the side of the windpipe.
2. Have the students locate one of their pulse points.
3. Ask the students to count the number of beats they feel while you measure 15 seconds with the timepiece.
4. Have the students multiply the number of beats by four. The answer will be their number of heartbeats in one minute, or their heart rate.

EXTENSION:
Help the students get a better idea of how a heart rate can vary. Give each student a copy of the worksheet on the following page and have them do the research in class or at home.

...fourteen, fifteen, sixteen, seventeen....

Heart Rate Worksheet

This worksheet will help you better understand how your heart beats in different situations. Count the number of times your heart beats in 15 seconds for each situation described in the worksheet. Then multiply this number by 4 to get your 60 second heart rate. You can use a watch or clock that has a second hand to time yourself.

Heart Rate Worksheet

Name: _____   Age: _____

Heart Rate Activity	15 second Heart Rate	60 second Heart Rate
Standing Heart Rate		
Sitting Heart Rate		
Lying Down Heart Rate		
Heart Rate After Running In Place For 1 Minute		
Heart Rate After Eating A Meal		
Heart Rate After Waking Up		
Heart Rate Before Sleeping		
Heart Rate While Walking		

PERSONAL PERCENTS

Knowing how much time we spend doing things, such as sleeping and being in school, helps us effectively manage our time.

MATERIALS:
Paper, pencil

DIRECTIONS:
1. Tell the class that each day can be divided into segments. These segments represent the time it takes to accomplish certain tasks.
2. Remind the students that there are 24 hours in one day. Have the class determine the number of hours they spend in school during one day. Students should write this number on their papers.
3. Have the students divide the number of hours they spend in school by the number of hours in one day.
4. Tell the class to multiply this answer by 100. This is the percent of the day they spend in school.

EXTENSION:
Have the students develop their understanding of time management with the Personal Percents Pie Charting activity on the following page.

Ten-Minute Real World Math ©1997 Monday Morning Books, Inc.

Personal Percents Pie Charting

Pie charts get their name because they are shaped like a pie, and information is shown as pieces of the pie.

Fill in the pie chart below by answering the following questions:

- What percentage of the day do you spend in school?
- What percentage of the day do you spend sleeping?
- What percentage of the day do you spend studying?
- What percentage of the day do you spend doing other things?

For example: If you spend 6 hours of the day in school, divide 6 by the number of hours in one day (6 divided by 24 = .25). Multiply .25 by 100 (.25 x 100 = 25%). 25% of your day is spent in school. On the pie chart, draw a piece of pie that is 25% of the whole pie. Label the piece "school."

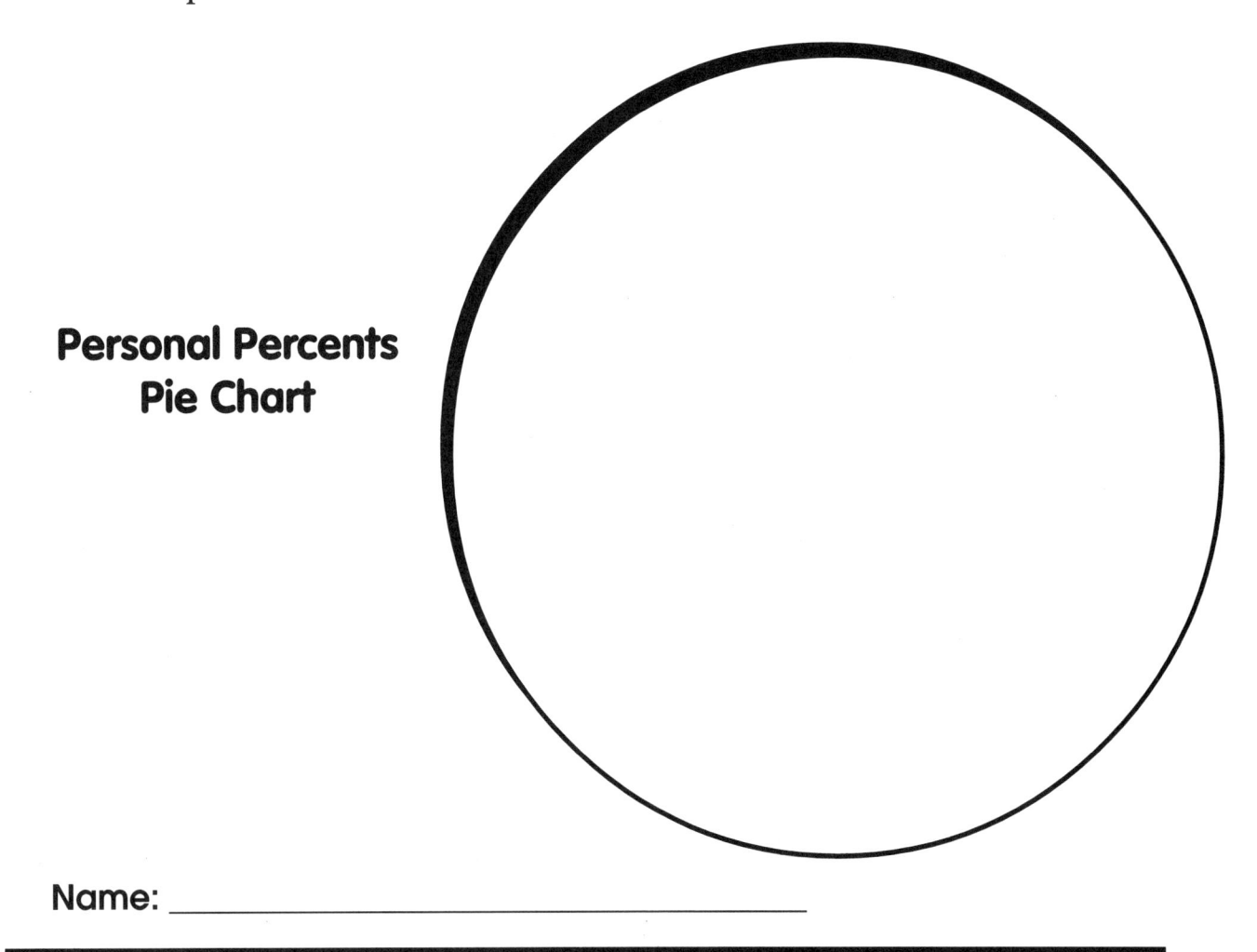

Personal Percents Pie Chart

Name: _____

HAIR COUNT ESTIMATING

Estimating a small portion of a group, and then applying the estimate to the whole group, is a useful method for quick calculations.

MATERIALS:
Paper, pencil, ruler for each group of two students

DIRECTIONS:
1. Tell the class they will be estimating the number of hairs on one of their forearms.
2. Have the class pair off into groups of two.
3. Tell one student to hold the ruler parallel to one forearm.
4. Have the other student count the number of hairs along one inch (2.5 centimeters) of the ruler. Students should write this number on their papers.
5. Now have the student hold the ruler across the forearm and count the number of hairs in one inch (2.5 cm). Students should write this number on their papers.
6. Have the students multiply the two counts together. This is the number of hairs in one square inch (2.5 square centimeters).
7. Tell the first student to measure the length and width of her or his forearm, and write down the results.
8. Have the students help each other multiply the forearm length by its width. This is the area of the forearm.
9. Tell the class to multiply the number of hairs in one square inch (2.5 sq. cm) by the area of the forearm. This is their estimate for the number of hairs on a forearm.
10. Tell the pairs to switch and repeat this estimate for the other student.

EXTENSION:
Challenge students to explore estimating by having them calculate the number of hairs on a family member's head. Remind them to first count the number of hairs in a small area, then apply this count to the whole area.

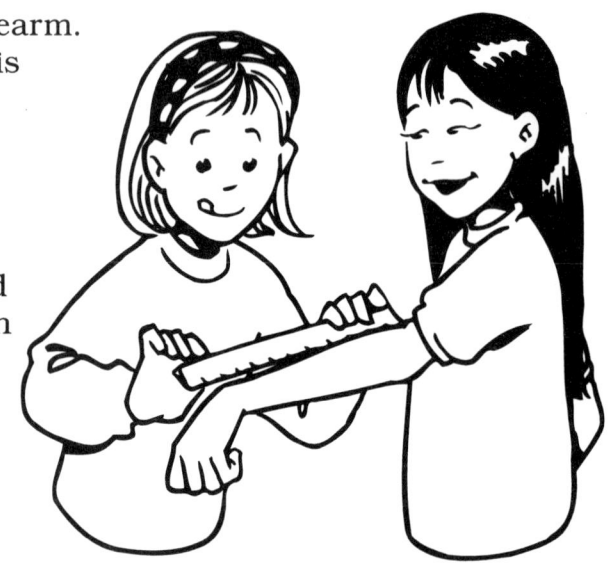

EYE BLINKS ESTIMATING

Eye blinking is an automatic reflex that helps keep the eye moist. Each time the eyelid closes a lubricant is spread out over the sclera. The sclera is the eye's outermost layer.

MATERIALS:
Paper, pencil, watch with second hand

DIRECTIONS:
1. Ask the students if they have ever wondered how many times their eyes blink in one hour.
2. Note the starting time and have the students begin silently counting each time their eyes blink. Students should blink at their normal rate.
3. After 30 seconds has elapsed tell the kids to stop counting. Students should write the number of eye blinks on their papers.
4. Have the students multiply the number of eye blinks by 2. This is their number of eye blinks in one minute.
5. Tell the class to multiply this number of eye blinks by 60. This is the number of times their eyes blink in one hour.

EXTENSION:
Reinforce multiplication skills, and students' eye awareness, by having the class calculate the number of eye blinks:
a) they have in one day while they are awake,
b) they have had in their lives, and
c) that Joan of Arc had in her lifetime.
Have the students present their results to the class.

Yes, and I lived for 19 adventure-filled years.

Are you really Joan of Arc?

Pace Counting Conversion

Converting a pace count into feet or meters is an excellent way to approximate a wide variety of measurements.

MATERIALS:
Tape measure, masking tape, paper, pencil

DIRECTIONS:
1. In preparation for this activity, measure 20 feet (7 meters) on the floor with a tape measure. This can be done in the classroom, hallway, or playground. Use tape to mark the beginning and end points of the 20 feet (7 meters).
2. Have students place their toes on the edge of the beginning tape mark.
3. Tell the students to step off with their right foot, and have them count each time their right foot hits the ground.
4. Instruct the students to stop counting when they reach the end tape mark.
5. Have students write their number of paces on a piece of paper.
6. To find the number of feet (or meters) in their pace, have the students divide the number of paces into 20 (7).

EXTENSION:
Pace counting is useful in identifying distances and areas. The activities on the following page will reinforce the students' knowledge of geometry, and give them confidence in their analytical skills.

Leg-Up Geometry

The area of a rectangle or square = Length times Width (A = L x W).
The perimeter of a rectangle or square = the sum of the length of all four sides (P = L + L + W + W).

What is the Area?

This activity requires a large rectangular or square area, such as a classroom, gymnasium, or parking lot. Divide the students into teams of two, with each team having a Length team member and a Width team member. Have the Length team members pace off the distance from one side of the chosen space to the other side of the space. Have the Width team members pace off the distance between the remaining two sides of the area. Ask the individual teams to regroup and convert their pace counts into linear measurements. Tell the teams to multiply their length measurement by their width measurement. Compare the various teams' answers for the space's area and have the students discuss reasons for differing answers.

What is the Perimeter?

This activity requires a large rectangular or square area, such as a classroom, gymnasium, or parking lot. Divide the students into teams of four, with each team having a Length-One member, a Length-Two member, a Width-One member, and a Width-Two member. Have each team member pace off one side of the area. Ask the individual teams to regroup and convert their pace counts into linear measurements. Tell the individual teams to sum their measurements. Compare the various teams' answers for the space's perimeter and have the students discuss reasons for differing answers.

SUN PROTECTION FACTORS

The Sun Protection Factor (SPF) is a rating used by sun screen manufacturers to guide consumers when buying sun protection products.

MATERIALS:
Paper, pencil

DIRECTIONS:
1. Tell the class that direct exposure to the sun's rays can be harmful. Wearing a hat, long sleeves, and long pants provide protection. Exposed skin should be protected with sun screen.
2. Explain that SPF numbers are used to calculate how many 15-minute time periods a sun screen's protection will last.
3. Write the following example on the board:

> SPF 8 means the sun screen should last for:
> 8 x 15 minutes = 120 minutes = 2 hours

4. Have the students calculate the number of minutes, and hours, of protection the following SPFs should provide: SPF 5, SPF 10, SPF 15, and SPF 20.
5. Review the answers with the class.
6. Tell the class that many health experts agree that SPF 15 is the minimum protection that people should use.

EXTENSION:
Have the students research and write a report on what the ozone layer is, and how this layer protects humans.

SCHOOL DAYS

Have the students ever wondered how many days they spend in school each year? This step-by-step activity will help answer this question.

MATERIALS:
Paper, pencil

DIRECTIONS:
1. Explain to the class they are going to estimate the number of days they are in school during one year.
2. Have the class write the number of days they are in school during one week.
3. Tell the students to multiply this answer by the number of weeks in one month. This is the number of days the students are in school during one month.
4. Have the class write down the number of months they are in school during one year.
5. Tell the students to multiply the number of school days in one month by the number of months they are in school during one year. The result is an estimate of the number of days a student spends in school during a year.

EXTENSION:
Further develop students' estimating skills by having them calculate: a) the number of days they have spent in school during their lives, b) the number of days they will spend in school by the time they graduate from high school, and c) the number of days they would spend in college.

TIME BUDGET

Scheduling time effectively is a skill that allows better planning and helps to determine how long a task should take.

MATERIALS:
Paper, pencil

DIRECTIONS:
1. Ask if any of the students have heard of a "time budget."
2. Explain that a time budget is a schedule of how time will be used. This schedule is made before the time being planned happens.
3. Have the class write down what time they arrived at school that day.
4. Have the students write down what time they will leave school that day.
5. Based on these times, have the class determine how many hours they will be in school that day. This is their school time budget for the day.

EXTENSION:

Help students develop time management skills by having them plan a time budget. Duplicate the Time Log activity on the following page and have the students plan the next day they will not be in school.

Time Log

Plan a time budget for a day that you are not in school. Use the chart below to plan your day before the actual day arrives. When the day arrives, see if you can use the budget as it is. It is okay if you have to make changes to your time budget, but be sure to mark these changes on the chart.

Time Log

Name: _____ Date: _____

Time	Planned Event	Actual Event
	Wake Up	

FAMILY PERCENTS

Calculating percentages is a method used to keep track of portions of a whole. Teaching students to successfully determine percentages will enable them to view events and objects as groups of smaller subsets.

MATERIALS:
Paper, pencil

DIRECTIONS:
1. Ask the class to write down the number of people they live with at home.
2. Have the class write down the number of people living at home that go to school.
3. Tell the students to divide the number of people in school by the number of people living at home.
4. Have the class multiply this number by 100. This is the percent of people at home who go to school.
5. Have the students review their answers with the class.

EXTENSION:
Reinforce students' understanding of percentages by having them complete the Family Percents Graphing activity on the following page.

Family Percents Graphing

Graph the answers to the following questions on the bar graph below:

- What percent of people at home go to school?
- What percent of people at home cook food?
- What percent of people at home watch television?
- What percent of people at home read books?

For example, if there are 6 people living at home, and 2 of them go school, divide 2 by 6 (2 divided by 6 = .33). Multiply this number by 100 (.33 x 100 = 33%). 33% of the people living at home go to school. You would then draw a bar on the graph below showing 33% of the people at home are in school.

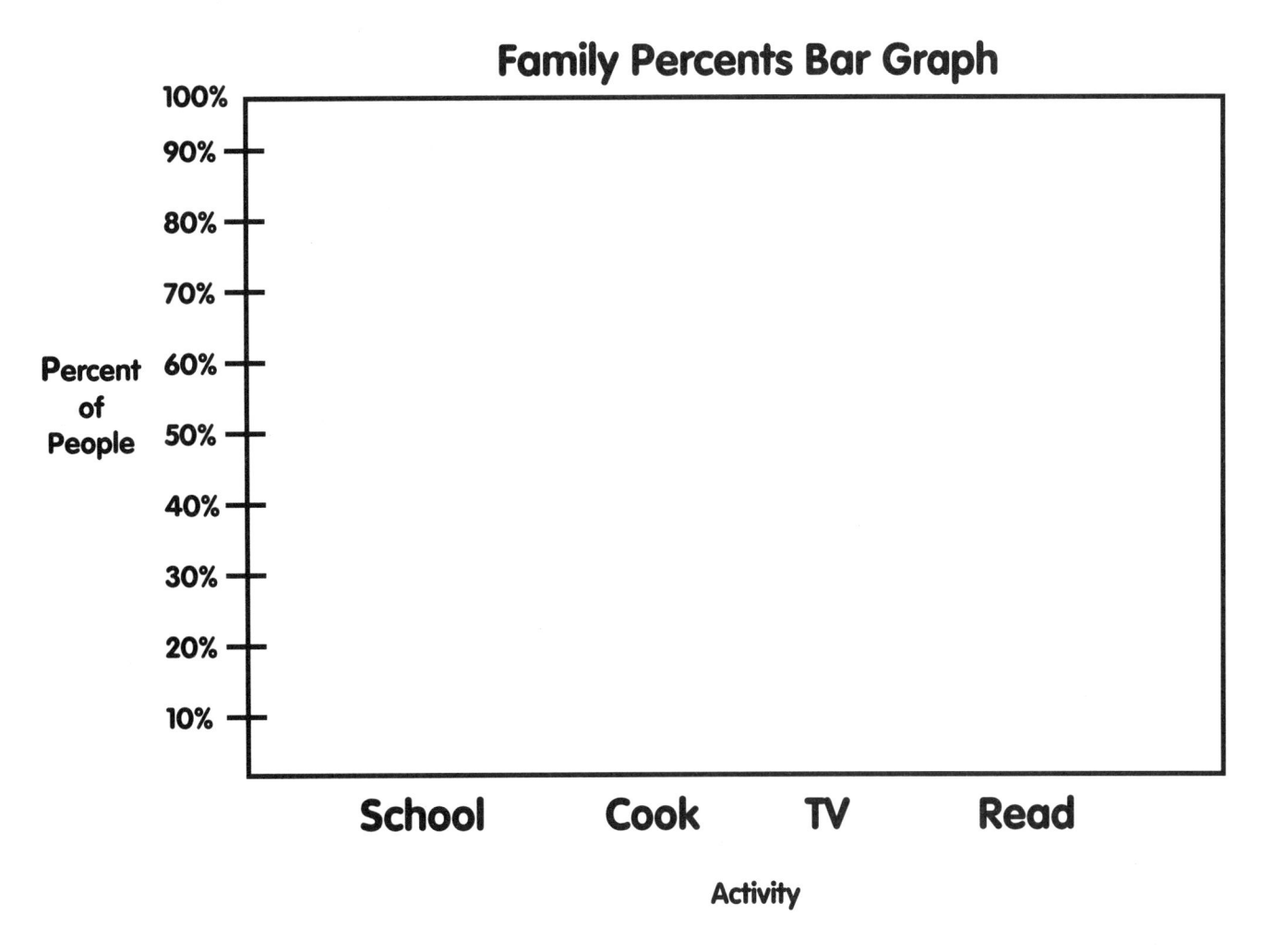

Family Percents Bar Graph

Percent of People (y-axis: 10% to 100%)

Activity (x-axis): School, Cook, TV, Read

RATE IT RIGHT

People often rate things on a scale from one to five. This way of expressing likes and dislikes is useful when evaluating a product or event.

MATERIALS:
Paper, pencil

DIRECTIONS:
1. Tell the class they will be rating how much they like math, on a scale of one to five. One means they don't like math at all, three means they think math is okay, and five means they like math a lot.
2. Ask the students to raise their hands if they give math a rating of five. Write this number on the board. Repeat this process for four, three, two, and one.
3. Review the results with the class. Ask the class why they like math and why they don't like math. (Be sure to add up the total number of ratings and see if everyone answered.)

EXTENSION:
Strengthen students' understanding of rating by having them complete the Simple Surveying activity on the following page. Students should present their survey result to the class.

Simple Surveying

Survey your family members, friends, or neighbors to find out how they rate the quality of shows on television. Use the worksheet below to keep track of your survey. Collect opinions from at least five people.

Television Quality Survey

Survey taken by: _____

Using the following rating scale, how do you rate the quality of shows on television?

Rating scale

5 = TV quality is excellent
4 = TV quality is good
3 = TV quality is fair
2 = TV quality is bad
1 = TV quality is very bad

Date	Person Surveyed	Rating

KID COUNT ESTIMATING

Estimating can be a quick way to approximate the answer to a mathematical question. Many times, estimates are used before more exact answers are calculated.

DIRECTIONS:
1. Explain to the class that they are going to estimate the number of students in their school.
2. Have each student silently count the number of kids in the classroom.
3. Ask the students to guess the number of classrooms in the school.
4. Have students multiply the number of students by the number of classrooms. This is the estimate of the number of kids in the school.
5. Optional: Have the students research the actual number of students in their school and compare this number to their estimates.

EXTENSION:
Using the same method, have the class estimate:
• the number of desks in their school
• the number of ceiling lights in their school
• the number of windows in their school

Ten-Minute Real World Math ©1997 Monday Morning Books, Inc.

HOW MANY BOOKS?

Estimating skills can be used to build students' confidence in their analytical ability and problem-solving techniques.

MATERIALS:
Paper, pencil

DIRECTIONS:
1. Explain to the class that they are going to estimate the number of books in the school's (or town's) library.
2. Ask the students to visualize one of the library's book stacks.
3. Have the students write down the number of shelves they think are in one of the book stacks.
4. Help the students estimate the number of books that are on one shelf of one stack.
5. Tell the class to multiply the number of books on one shelf by the number of shelves in one stack.
6. Have the students write down the number of stacks they think are in the library.
7. Ask the class to multiply the number of stacks in the library with the answer from step 5. This is their estimate for the number of books in the library.

EXTENSION:
Show the class how to develop a more accurate estimate by taking them to the library and having them count the number of stacks. Also, tell them to count the number of shelves in one stack and the number of books on one shelf. Help the students multiply their book, shelf, and stack counts to estimate the number of books in the library. Ask the librarian to verify the accuracy of their estimates.

COUNTING HAIKU

Haiku poems originated in Japan and are considered a compact form for expressing ideas and feelings.

DIRECTIONS:
1. Explain to the class that Haiku are poems that have three non-rhyming lines. The first line has exactly five syllables; the second line has exactly seven syllables; and the third line has five syllables.
2. Tell the students that one way of composing haiku is to write three sentences about something without counting the number of syllables. Then trim the sentences to make a haiku.
3. Write the following on the board:

> The icy cold sting of mountain air
> The sound of crunching footsteps in the fresh snow
> Reaching up to touch the sky

4. Have the students count the number of syllables in each sentence.
5. Rewrite the sentences as shown:

> Icy mountain air
> Crunching footsteps in fresh snow
> Reaching to touch sky

6. Ask the class to count the number of syllables in each sentence.

EXTENSION:
For homework, have the students write their own haiku. Then have the students recite their work in class.

How many syllables are in a haiku?

MAP READING MATH

Successful navigation skills involve graphic interpretation, arithmetic, and reading.

DIRECTIONS:
1. Draw the following road map segment on the board. Indicate what the numbers stand for in miles or kilometers.

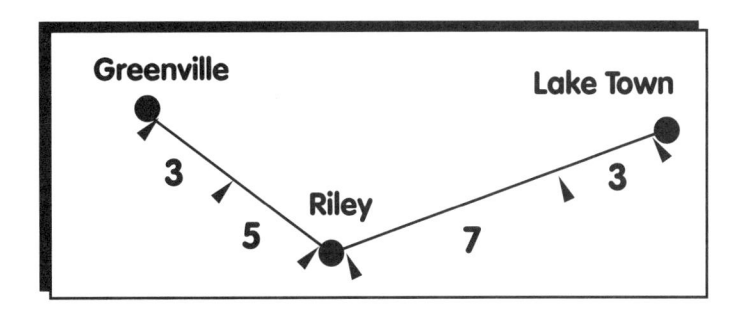

2. Explain to the students that many road maps show the number of miles (kilometers) of roadway. Often, these miles are listed between arrow points.
3. Ask the class to calculate the distance between Greenville and Lake Town. (18)
4. Have the students calculate how many minutes it will take to travel from Greenville to Lake Town if the car they are in is going 45 miles (kilometers) an hour. (18/45 = .4 hours, .4h x 60min/h = 24 minutes)

EXTENSION:
Have the students calculate the number of miles (kilometers) from their hometown to the nearest college or university. Also, have them estimate how long it will take to get there if the car goes 55 miles (kilometers) an hour. Travel maps and roads maps can be found in libraries or at travel agencies.

HOW FAR?

Have your students ever wondered how many trips they have made to the cafeteria? This activity will help them calculate their number of trips and the total distance they've travelled. (Note: if your school doesn't have a cafeteria, use the library, playground, or gymnasium.)

MATERIALS:
Paper, pencil

DIRECTIONS:
1. In preparation for this activity, have the class do the Pace Counting Conversion activity (p.40).
2. Have the students walk to the cafeteria, counting their paces as they go. Students should write the pace count on their papers.
3. Have the class walk from the cafeteria to the classroom, counting their paces as they go. Students should write the pace count on their papers.
4. Tell the class to sum the two pace counts and divide by two. This is the average pace count between the classroom and the cafeteria.
5. Have the students convert their pace counts into feet (meters).
6. Have the class multiply this result by the number of school days in a week, and then by the number of weeks in a school year. This answer is the distance they travel to the cafeteria in one school year.

EXTENSION:
For homework, have the students calculate: a) the distance they travel from their bedrooms to their kitchens in one year, and b) the distance they have travelled from their bedrooms to their kitchens in their lifetime.

If I go from my bedroom to the kitchen seven times a day,....

DRAWING WITH SHAPES

Many artists use geometry when they begin a work. By using mathematical shapes, the artist can approximate how the finished work will look.

MATERIALS:
Paper, pencil, ruler

DIRECTIONS:
1. Tell the students they will be drawing with shapes. Remind the students that an equilateral triangle has three sides of equal length. Have each student draw an upside-down equilateral triangle.
2. Have the students draw a circle centered on the top side of the triangle. They can draw two eyes, a nose, and a mouth on the circle.
3. Tell the class to draw thin rectangles coming down from each top corner of the triangle.
4. Ask the students to draw a square centered on the lower tip of the triangle.
5. Have the students draw a rectangle coming off each lower corner of the square.
6. Students can add hands and feet to their drawings.

EXTENSION:
Have students develop their creativity, and strengthen their understanding of geometry, by drawing a geometric cartoon strip. Their strips should include words that tell a short story.

DRAWING TO SCALE

Drawing something to scale means making the dimensions of the picture proportionally match the actual object. Understanding scale is useful in many areas, including: map reading, drawing, and graphic design.

MATERIALS:
Paper, pencil, ruler (for each student)

DIRECTIONS:
1. Explain to the class that they are going to learn how to draw a scale picture of a building.
2. On the board, draw a building and label it 24 feet (8 meters) tall and 30 feet (9 meters) wide.
The scale of the drawing will be:
- 1/4 inch equals one foot
- 2 centimeters equals 1 meter

Have the students write the scale they will use on their papers.
3. Ask the class how many inches (cm) tall the building should be drawn if they draw it to scale. Have them look at their rulers and count each quarter inch (2 cm) off as a foot (meter). The building will be 6" (16 cm) tall.
4. Have the class measure the height with their rulers and draw one side of the building on their papers.
5. Ask the class how wide the building should be if it is drawn to scale. Have the students count off each quarter inch (2 cm) as a foot (meter) on their rulers. The building will be 7 1/2" (18 cm) wide.
6. Have the class draw the bottom of the building to scale on their paper.
7. Tell the students to finish their scale building by drawing the other side and the top of the building.

EXTENSION:
Reinforce the importance of understanding scale by having the class draw a row of three foot by six foot (one meter by two meters) scale windows on their buildings.

ESTIMATING DISTANCE

Gauging distance is easily done when a smaller, standard distance is known. Estimating distance is a useful skill when an approximate measure is all that is needed.

MATERIALS:
Paper, pencil, measuring tape

DIRECTIONS:
1. Tell the class that they are going to measure how tall they are, and then use that measurement to estimate the height of the doorway.
2. Extend the tape measure and have each student come up and measure his or her height. Students should write their heights on their papers.
3. After a student gets measured, the student should stand in the doorway. Have the student touch the edge of the doorway where the top of his or her head touches the edge.
4. Still touching the doorway, the student should step back and estimate how many more body lengths would reach the top of the doorway. Students should write this number on their papers.
5. Have the class covert the number of body lengths of doorway height into a standard length using their known body height measurement. (Multiply the number of body lengths by the height measurement.) This answer is the estimate of the doorway height.

EXTENSION:
Estimating skills can be strengthened by having the students use their body height to measure: a) the length of their beds, b) the height of a table, and c) the width of a doorway. Have the students present their findings to the class.

This table is one-third of my height.

DIVIDE BY EYE

Mathematicians, scientists, and engineers often use an easy-to-follow rule when estimating a measurement: The smallest increment of measure on a measuring device (ruler, gauge, etc.) can be halved by eye. For example: if you are taking a measurement with a ruler, and the smallest increment on the ruler is $1/4$ inch (.5 centimeters), you can accurately estimate one-half of $1/4$ inch (.5 cm) by visually dividing $1/4$ (.5 cm) in half.

MATERIALS:
Paper, pencil

DIRECTIONS:
1. Tell the class they are going to learn a useful method for estimating the smallest unit of measure.
2. Draw the following portion of a ruler on the board:

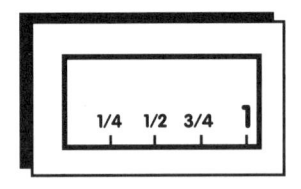

3. Explain to the students they can estimate the smallest increment of this ruler by eye. The smallest unit of measure marked on this ruler is $1/4"$.
4. Students should look at the halfway point between the 0" mark and the $1/4"$ mark. Ask the class what this measurement is. ($1/8"$)

EXTENSION:
Help students develop and strengthen this skill by having them estimate the smallest units of measure on various measuring devices, such as a gas gauge in a car, a tape measure, a thermometer, a clock, and a protractor. Students should present their findings to the class, including: a) the name of the measuring device, b) the smallest unit of measure marked on the device, and c) the estimate of the smallest unit when dividing by eye.

BALLPARK GEOMETRY

An interesting and useful method of estimating is called "Ballparking." This style of approximation uses known geometric sizes, such as a football field, to estimate size or distance.

MATERIALS:
Paper, pencil

DIRECTIONS:
1. Ask the students to look around the room and notice how big the classroom is.
2. Have the class visualize another, larger room in the school, such as the library, the gymnasium, or cafeteria.
3. Have the students estimate how many "classroom lengths" long this larger room is.
4. Students will have varying answers. Discuss possible reasons for differing answers with the class.

EXTENSION:
Help the students hone estimating skills by verifying the approximation they did in the exercise above. Measure the length of the classroom with a tape measure, then take the class to the larger room and have them measure its length. Ask the students to compare their ballpark estimates with the actual measured dimensions.

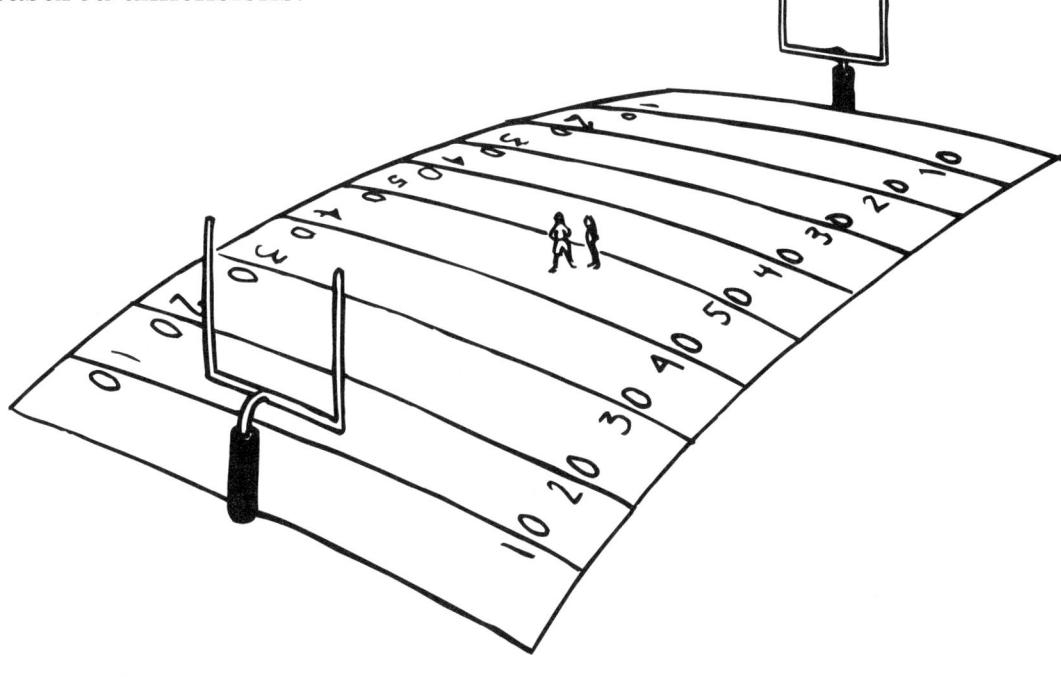

MEASURING WITH PAPER

Often a ruler is not available when a measurement needs to be taken. A piece of paper, with known dimensions, can be used to obtain accurate results.

MATERIALS:
Paper with known dimensions, a pencil, a ruler

DIRECTIONS:
1. Explain to the class that they will be making a measuring device.
2. Give each student a piece of paper and tell the class the length of the long side of the paper.
3. Have the class fold the long side of their papers three times, as shown in the diagrams.
4. Have the students unfold their papers.
5. Help the class mark the folds on their papers as follows:

- the middle fold is 1/2 the length of the paper's long side
- fold **b** is 1/4 of the length
- fold **e** is 3/4 of the length
- fold **a** is 1/8 of the length
- fold **f** is 7/8 of the length
- fold **c** is 3/8 of the length
- fold **d** is 5/8 of the length

6. Have the students measure various things in the classroom, such as the length and width of their desks, the length of a pencil, the width of a book, and the length of the chalkboard.
7. Students can verify their paper ruler measurements by measuring the items in step 6 with a standard ruler.

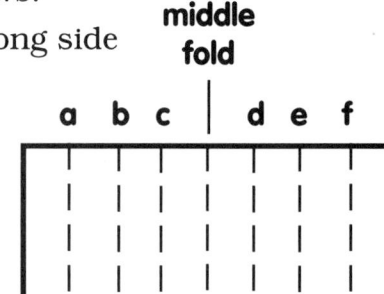

EXTENSION:
Being able to improvise measurements is an ingenious skill that kids can use throughout their lives. For homework, have the students brainstorm a different way to make a measuring device. Then ask them to bring these devices to class and explain how they work.

TOOL-OF-THUMB SCALING

Artists sometime use their thumbs to draw to scale. Holding the thumb up to an object makes it easier to draw the object in correct proportion.

MATERIALS:
Paper, pencil

DIRECTIONS:
1. Tell the class they are going to learn how to use their thumbs for drawing to scale.
2. Have each student close one eye and hold a thumb up while looking at the room's doorway.
3. Tell the class to measure how many "thumbs high" the doorway is. Students should write this number on their papers.
4. Now have the students hold their thumbs next to their papers, and have them draw the edge of the doorway. Remind the class to measure this edge by the number of "thumbs high" the doorway is.
5. Tell the class to close one eye, hold up a thumb, and notice how many "thumbs wide" the doorway is. Students should write this number on their papers.
6. Have the students draw the top width of the doorway using the "thumbs wide" measurement.
7. Have the students complete their doorway drawings by drawing the other edge and bottom width.

EXTENSION:
Challenge students' use of perspective and proportion by having them draw the outside of a familiar building, such as their school, their homes, or their local store. Have the students share their thumb-scale drawings with the class.

FINGERTIP MEASURING

For centuries, people have used outstretched arms to measure length. Knowing the distance between your fingertips can be a useful aid in measuring without a ruler.

MATERIALS:
Measuring tape, paper, pencil

DIRECTIONS:
1. Have two students come to the front of the class.
2. Give the end of the tape to one student, and have the other student pull out about six feet (2 meters) of tape.
3. While these two students are holding the tape, have their classmates come up one at a time, extend their arms to their sides, and measure the length between their right fingertips and left fingertips.
4. Tell the students to write this measurement on a piece of paper when they return to their seats.
5. Have the class measure, with their fingertips, various items in the classroom, such as the length of a wall, the width of the doorway, and the height of a desk. They should record these measurements on a piece of paper. (Measurements of less than one length can be estimated as a quarter length, one-half a length, or three-quarters of a length.)
6. Help the students use their distance between fingertips measurement to convert the number of fingertip lengths into a standard unit of measure. (Multiply the number of fingertip lengths by the known distance between fingertips.)
7. The class can verify the accuracy of fingertip measuring by measuring the same items with the tape measure.

EXTENSION:
Duplicate enough copies the Fingertip Measuring Chart, on the following page, for each student. For homework, have the students use the chart to reinforce the concept of fingertip measuring and conversion.

Fingertip Measuring Chart

Use your fingertip measuring skills to determine the measurements of the items on the Fingertip Measuring Chart.

Fingertip Measuring Chart

Name: _____

The distance between my fingertips is: _____

Item to Measure	Fingertip Width	Converted Width
Length of Bed		
Width of Bed		
Height of Bed		
Refrigerator Height		
Refrigerator Width		
Sink Height		
Sink Width		
Room Length		
Room Width		
Table Length		
Table Width		
Chair Height		
Chair Width		

Change fingertip width into converted width by multiplying the fingertip width by the distance between your fingertips.

SYLLABLE COUNTING

Syllables are parts of words that form a continuous, uninterrupted sound when spoken. Breaking words into their syllables is a useful tool for mastering meaning, spelling, and phonetic pronunciation.

MATERIALS:
Paper, pencil

DIRECTIONS:
1. Remind the class that syllables are pronunciation units. Some words have only one syllable (math), and some words have many syllables (mathematics).
2. Write the following sentence on the board:

> Sixteen students rode the bus to school.

3. Have the students write the sentence on their papers. Students should count the number of syllables in each word.

> (2)   (2)   (1) (1) (1) (1) (1)
> Six-teen stu-dents rode the bus to school.

4. Have the class sum up the number of syllables in the sentence.

EXTENSION:
Teach students how to calculate what grade level a book is written for by completing the Calculating Readability activity, on the next page, with the class.

CALCULATING READABILITY

Students' ability to read at a given grade level can be verified using the Fog Index. This readability gauge, developed by Robert Gunning, calculates the level of education needed to read a written piece.

MATERIALS:
Paper, pencil, written text with at least 100 words (novel, textbook, newspaper article, etc.)

DIRECTIONS:
1. Explain to the class that they are going to learn how to calculate the grade level something is written for. The Fog Index is a four-step method for determining the education level of written work.
2. Have the students do the following:
• Determine the average sentence length. Count the number of words on a page and the number of sentences on a page. Divide the total number of words by the number of sentences. This answer is the average sentence length.
• Determine the percent of hard words. A hard word is defined as one that has at least three syllables. Count the number of hard words on a page. Divide the number of hard words by the number of total words. Multiply this answer by 100. This is the percent of hard words.
• Add the average sentence length to the percent of hard words.
• Multiply this result by .4. This answer is the grade level needed to read the written work.

For example: 1) This page has 210 words and 26 sentences, for an average sentence length of 8. 2) The number of hard words on this page is 27. The percent of hard words = (27/210) x 100 = 13. 3) 8 + 13 = 21. 4) 21 x .4 = 8.4. This page can be read by someone who reads at between an eighth and ninth grade level.

EXTENSION:
Have students do a readability calculation as part of their next book report.

24-HOUR TIME

Many countries, agencies, and people use the 24-hour clock to tell time. Being able to tell 24-hour time enables students to: a) understand Greenwich Mean Time, b) make a watch-compass, and c) communicate the time in another way.

DIRECTIONS:
1. Write the whole numbers from 1 to 12 on the board, as shown below.

1AM	7AM	1PM	7PM
2AM	8AM	2PM	8PM
3AM	9AM	3PM	9PM
4AM	10AM	4PM	10PM
5AM	11AM	5PM	11PM
6AM	12PM	6PM	12AM

2. Explain that 1 AM through 12 PM (noon) is represented by 0100 through 1200. Reinforce this by writing the 24-hour times next to their AM equivalents. (See the illustration below.)
3. Explain that 1 PM through 12 AM (midnight) is represented by 1300 through 2400. Reinforce this by writing the 24-hour times next to their PM equivalents.
4. Show the students an easy way to covert 12-hour time to the 24-hour clock. For times after 1 PM, add twelve to the number. For example, 2 PM becomes 14. Multiply this number by 100. (2 PM equals 1400, pronounced: fourteen-hundred.)
5. Demonstrate to the students that fractions of an hour are placed on the end of 24-hour time. For example, fifteen minutes past six in the evening equals 1815. (This is pronounced eighteen-fifteen.)
6. Explain to the class that one minute past midnight through 1 AM are represented by 0001 through 0100. Forty-eight minutes past midnight equals 0048.

EXTENSION:
Show your students how exciting it is to turn a simple, dial-faced watch into a compass, using the exercise on the following page.

Ten-Minute Real World Math ©1997 Monday Morning Books, Inc.

It's a Watch . . . It's a Compass . . .

Never be lost again! A round watch with a dial face (analog watch) can be turned into a compass.

- Read the time on the watch in 24-hour time, and divide this number by two.
- Looking at the watch, point the answer's time in the direction of the sun.
- Without moving the watch, look in the direction of 1200 on the watch. You are now looking to the north. West is to your left, east is on the right, and south is behind you.
- For example, you look at your watch, and it reads 1400. Half of 1400 is 0700. Point 0700 in the direction of the sun. The position 1200 on the watch is pointing to the north.

That way is north!

GREENWICH MEAN TIME (GMT)

The prime meridian (0 degrees longitude) runs through Greenwich, England, and is the starting point for the world's time. GMT is based on the 24-hour clock and all other time zones are established by converting from GMT.

DIRECTIONS:
1. Review the 24-Hour Time (p. 66) activity with the class.
2. Explain to the students that the current time is actually based on what the time is in Greenwich, England. Local time is calculated by adding or subtracting hours from GMT.
3. Have the class convert the following local 24-hour times to GMT by adding or subtracting the appropriate number of hours:

Place	Local Time	GMT
San Francisco, USA	0900	+8 hours
New York, USA	1515	+5 hours
Sydney, Australia	2200	-10 hours
Yellowknife, Canada	1300	+7 hours
Paris, France	0500	-1 hour

EXTENSION:
To further strengthen the students' understanding of Greenwich Mean Time, have them research and report on what time they were born in GMT.

WEATHER MAP DATA

An important piece of information given on weather maps is wind speed. High winds can be extremely dangerous.

MATERIALS:
Paper, pencil

DIRECTIONS:
1. Explain to the students that wind speed is sometimes given in terms of knots, instead of miles per hour. A hurricane is defined as a storm with winds of 64 knots or more.
2. Write the following on the board:

> 1 knot = 1.15 miles per hour (1.85 kilometers per hour)

3. Help the students understand the power of hurricane winds by having them convert 64 knots to its equivalent in mph (kph). (73.6 mph, 118.4 kph)

EXTENSION:
Develop students' understanding of wind by having them research and report on the El Nino wind phenomenon. When they present their findings to the class, they should include any wind speed conversions necessary.

CENSUS REPORT

Every ten years, the U.S. Government counts the number of people in the country. This Census information is used to chart population growth.

MATERIALS:
Paper, pencil

DIRECTIONS:
1. Explain to the class that a census is a type of survey that counts the number of people in a country.
2. Demonstrate how a census works. Have all the girls in class raise their hands and have one boy count the number of girls. Write this number on the board.
3. Have the boys raise their hands and have one girl count the number of boys in the class. Write this number on the board.
4. Have the students calculate the total number of students in the class by adding the two numbers.
5. Have the students determine the percentage of girls and boys in the class.
(The percentage of boys: (number of boys/total in class) x 100, percentage of girls: (number of girls/total in class) x 100.)

EXTENSION:
Strengthen students' understanding of statistics by having them research and report on population changes in their country. Their reports might include data from years past, and projections for population growth into the next century.

Every ten years, the Census counts the number of people.

TEMPERATURE TRACKING

Two important meteorological measurements are the daily high and low temperature readings. A useful piece of information is the difference between high and low temperatures. Knowing this difference can influence a variety of decisions, including how people dress, how farmers care for crops and livestock, and how athletes train.

MATERIALS:
Thermometer, watch, paper, pencil

DIRECTIONS:
1. Explain to the students the difference between high and low temperatures is called the temperature range.
2. At the beginning of the school day, place the thermometer outside. After 15 minutes elapses, have the students read and record the temperature, and write down the time.
3. Halfway through the day, have the class go back to the thermometer to read and record the temperature and time.
4. Have the class subtract the morning temperature from the afternoon temperature. This is the change in temperature.
5. Have the students calculate the difference in time by subtracting the morning time from the afternoon time.
6. Ask the class to analyze the information: Did the temperature go up or down? If so, why do they think it changed?

EXTENSION:
Have the class calculate the change in temperature per hour. This is done by dividing the change in temperature (degrees) by the change in time (hours). For an additional challenge, students can study temperature changes over an extended period of time.

MARS MISSION MATH

The fourth planet from our sun is Mars, also known as the red planet. Several robotic spacecraft have journeyed to Mars, and research is underway to meet the challenges of putting humans on this distant planet.

DIRECTIONS:
1. Tell the students they will calculate the amount of food and water needed for a trip to Mars.
2. On the board, write the following information:

duration of trip to Mars =	310 days
food per astronaut/day =	1 pound (.5 kg)
water per astronaut/day =	1/2 gallon (2 liters)
1 gallon (4 liters) of water =	8 pounds of water (4 kg)

3. Have the students calculate the total number of pounds (kilograms) of food and water needed for three people to travel to Mars and back.

Answer:
(2 x 310 days)(1 lb food per day x 3)
= 1,860 lb of food
(2 x 310 days)(1/2 gal water per day x 3)(8 lb water/1 gal water)
= 7, 440 lb water

Total weight of food and water = 1,860 lb + 7,440 lb = 9,300 lb

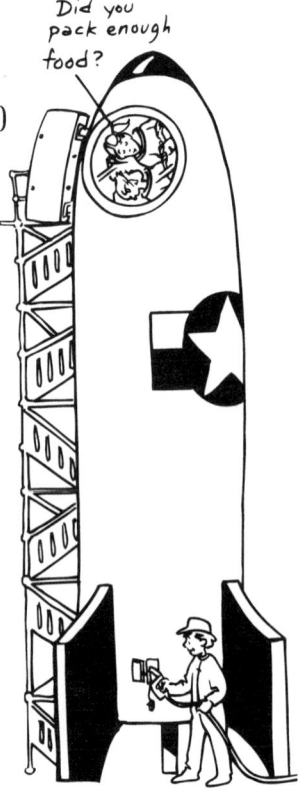

EXTENSION:
For More Martian Math fun, either in class or for homework, have students explore the exercises on the following page.

Answers to next page:
1. 6.8 sols. 2. 301.2 sols per one-way trip.
3. 17,750 pounds (8,875 kg) food and water.

More Martian Math

Use the following data to solve the problems below:

A Martian day is called a sol.
One sol equals 24.7 Earth hours.
An astronaut eats 1 pound (.5 kilograms) of food per day.
An astronaut drinks 1/2 gallon (2 liters) of water per day.
It takes 310 Earth days to travel to Mars.
One gallon (4 liters) of water weighs eight pounds (4 kilograms).

Problems:

1. How many sols are there in one Earth week?
2. How many sols are there in a one-way voyage to Mars?
3. If five astronauts travel to Mars, work on building a space base on the planet for ninety days, and then travel back to Earth, how many pounds (kilograms) of food and water will they have to bring?

GLOBAL POSITIONING SYSTEM

The Global Positioning System (GPS) is a set of satellites that orbit Earth. GPS is used to locate any point on the globe.

MATERIALS:
Pencil, paper

DIRECTIONS:
1. Explain to the class that the Global Positioning System uses orbiting satellites to help airplanes, ships at sea, and people locate their exact position.
2. Demonstrate that geometry can be used to establish position by drawing the following on the board:

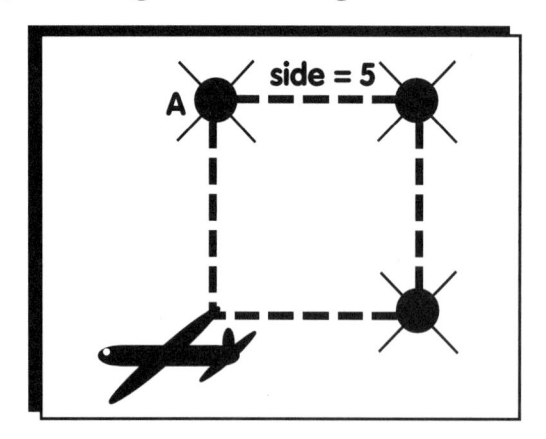

3. Tell the class that the three satellites and the plane are four corners of a square.
4. Explain that if the distances between the satellites are known, the fact that a square has equal length sides can be used to locate the distance of the plane from the satellites.
5. Ask the students how far the plane is from satellite A. (Answer = 5 miles or kilometers.)

EXTENSION:
Strengthen students' geometry skills and spacial concepts with the Shape Locators game on the following page.

Shape Locators

You and a friend can learn about geometry with this Shape Locators game. Gather the following materials before starting:

- two pencils
- two pieces of paper
- one ruler

1. Decide which one of you will be the "Satellite" and which one will be the "Airplane." The Satellite will be giving directions so that the Airplane can locate itself.

2. The Airplane should get ready with the ruler, a pencil, and a piece of paper.

3. The Satellite tells the Airplane to draw a rectangle. The long sides of this rectangle will be six inches. The short sides will be four inches. The Satellite must tell the Airplane to label the top left corner of the rectangle with an "A", the bottom left corner with a "B", the top right corner with a "C", and the bottom right corner with a "D."

4. The Satellite now tells the Airplane that the Satellite has spotted the Airplane flying on the line between "A" and "C", and the Airplane is halfway between "A" and "C."

5. The Airplane must calculate how far it is from "A" and tell the Satellite the answer.

6. Now the Satellite tells the plane it is on the line between "A" and "B", and is halfway between "A" and "B."

7. The Airplane must calculate how far it is from "A" and tell the Satellite the answer.

8. Start the game over, this time switching who is the Satellite and who is the Airplane.

After both team members have been the Airplane start over again. Choose a different size rectangle. Vary the game by having the Airplane be different distances from the corners.

METRIC CONVERSIONS

There are basically two systems of measurement in the world, metric and English. Each system has standards for distance, weight (mass), and time. In the English system, these are called pounds, feet, and seconds. The metric system, also known as the International System, calls them meters, grams, and seconds.

MATERIALS:
Paper, pencil, students' heights from Estimating Distance activity (p. 57)

DIRECTIONS:
1. Explain to the class that there are two systems of measurement, the metric and English systems.
2. Tell the students the unit of distance is called the meter in the metric system, and the foot in the English system.
3. Write the following conversion on the board:

> 1 meter = 3.2 feet

4. Have the students convert their heights from one system to the other system. If the measurement was taken in the metric system, multiply the height in meters by 3.2 to get feet. If the measurement was taken in the English system, divide the height by 3.2 to get meters. Remember that the heights must be in meters or feet. This means inches and centimeters must be changed to feet and meters.

EXTENSION:
Strengthen students' understanding of the metric and English systems by having them measure how much they weigh in one system and convert it to the other. Remind the class that 1 pound = 2,204 grams.

Conversion Basics
1 m = 3.2 ft. Divide both sides by 3.2 ft. This operation results in 1 m/3.2 ft = 1. This conversion factor can be used to convert meters to feet, or to convert feet to meters. Simply arrange the fraction so that the units desired are in the numerator, and multiply this fraction by the number being converted. For example, converting 6 meters to feet:
6m x (3.2ft / 1m) = 19.2ft.

One meter equals 3.2 feet.

COMPUTER MEMORY MATH

Knowledge of computers is becoming more essential in modern times. The space available on a computer, its memory, is important to know for software usage, graphic design, and games.

MATERIALS:
Paper, pencil

DIRECTIONS:
1. Explain to the class that computer memory is the amount of space available for storage on the computer. Most computers made today have a minimum memory of 500 megabytes. A megabyte is one million bytes. A byte is the unit that stores one character, such as a letter or single digit number.
2. An average page has about 2000 characters on it. Have the class calculate the number of bytes needed to store five pages. (Answer = 2000 bytes per page x 5 pages = 10,000 bytes)
3. Now have the students determine the number of pages that a 500 megabyte computer can hold. (500,000,000 bytes x (1 page/2000 bytes) = 250,000 pages)

EXTENSION:
Some new computers have a gigabyte of memory. One gigabyte equals one billion bytes. Have the students calculate how many pages could be stored on this type of computer.

SPORT STATISTICS

A sports team's progress can be tracked through their seasonal statistics. These statistics are calculated using basic math skills, such as addition, subtraction, multiplication, and division.

MATERIALS:
Paper, pencil

DIRECTIONS:
1. Tell the students that a sports team's performance is told in the form of statistics.
2. Write the following on the board, and have the class copy it onto their papers:

The Comets basketball team	
<u>wins</u>	<u>losses</u>
30	12

3. Have the class calculate the total number of games the Comets played. (42)
4. Have the students calculate the Comets' percentage of wins: (30/42) x 100 = 71.4%.
5. Have the students calculate the Comets' win-loss ratio: 30/12 = 2.5.

EXTENSION:
Develop students' basic math skills by having them complete the Team Tracking activity on the next page. Help the class get started by reviewing the sports statistics section of a newspaper.

I usually make 70% of these shots.

Team Tracking

Choose three favorite sports teams in one sport and use the chart below to track their progress for one month. The sports section of a newspaper should have all the information you need.

Team Tracking Chart

Name: _____
Team A: _____
Team B: _____
Team C: _____

	Wins	Losses	Ties
Team A			
Team B			
Team C			
	% Wins	**% Losses**	**% Ties**
Team A			
Team B			
Team C			

Team With Most Wins: _____
Team With Most Losses: _____

To calculate percent wins, divide the number of wins by the total number of games. Multiply this answer by 100 to get the percent of wins. Percent losses and percent ties can be determined the same way.

EURO DOLLARS

Many countries in Europe have formed the European Union (EU). One of the EU's goals is to have a standard currency that can be used in all the EU countries. This new, multinational currency is called the European Currency Unit, or the Euro.

MATERIALS:
Paper, pencil

DIRECTIONS:
1. Explain that money from one country is converted to money from another country by using an exchange rate. Because one nation's currency might be worth more, or less, than another nation's currency, the exchange rate is used to make the two different currencies worth the same.
2. Write the following on the board:

The countries of Europe will have one currency, the Euro.

> 1 Euro = 2 Dollars

3. Have the class calculate the exchange rate to convert from Dollars to Euros. (Answer: 1 Euro/2 Dollars = 0.5 Euros per Dollar. To exchange Dollars, multiply them by .5.)

EXTENSION:
Strengthen students' understanding of the international monetary market by having them complete the Currency Exchange activity on the following page. Remind the class that exchange rates can be found at libraries, banks, travel agencies, and the business or financial section of the newspaper.

Conversion Basics
1 Euro = 2 dollars. Divide both sides by 2 Dollars. This operation results in
1 Euro/2 Dollars = 1.
This conversion factor can be used to convert Euros to Dollars, or to convert Dollars to Euros. Simply arrange the fraction so that the units desired are in the numerator, and multiply this fraction by the number being converted. For example,
converting 6 Dollars to Euros:
6 Dollars x (1 Euro/2 Dollars) = 3 Euros.

Currency Exchange

Use the chart below to convert one country's money to Dollars. Currency exchange rates can be found at libraries, banks, travel agencies, and in the business or financial section of the newspaper.

Exchange Rates

Currency	Rate	100 Dollars' Worth
Austrian schilling		
English pound		
French franc		
Georgian lari		
German mark		
Indian rupee		
Italian lira		
Japanese yen		
Mexican peso		
Russian rouble		
Swedish krona		
Swiss franc		

How many French francs can I get for 100 dollars?

PAYING FOR TELEVISION

People starring in popular television shows tend to get paid well. A 30 minute show is generally 23 minutes long, with 7 minutes of advertising. How much should advertisers be charged in order to pay the cast's salaries?

MATERIALS:
Paper, pencil

DIRECTIONS:
1. Explain to the class that television shows are paid for with advertising revenue. Students will calculate the amount of money charged for ads, assuming ads pay only for the cast's salaries.
2. A successful TV show pays a four member cast 200,000 dollars per person for each 30 minute show. Have the students calculate the total salary cost of one show.
(Answer = 800,000 dollars)
3. Seven minutes are allotted for commercials during each show. If each commercial is one minute long, have the students calculate the cost of each commercial.
(Answer = 114, 285.71 dollars)

EXTENSION:
Have the students: a) calculate the quantity of 89 cent hamburgers an advertiser needs to sell to pay for a one-minute ad, b) the quantity of 20,000 dollar cars an advertiser must sell, and c) how much each commercial would cost if there were 14 ads that were 30 seconds each.

MATHEMATICIAN'S GLOSSARY

Acute Angle
An angle less than 90 degrees.

Algorithm
A procedure or method used for calculating. An example is dividing the numerator by the denominator and multiplying by 100 to calculate a percent.

Angle
The shape formed by two lines when each line starts at the same point.

Area
The portion of a shape surrounded by a line or lines. For example, the portion of a circle that is inside the circle's line.

Arithmetic
Problem solving that involves addition, subtraction, multiplication, and division.

Bar Graph
A drawing that uses lengths of parallel rectangles to display information.

Chart
A graph or a table used to present information.

Circumference
The boundary line of a circle.

Decimal
A number that represents a fraction, for example, the fraction 1/10 is represented by the decimal 0.1.

Mathematician's Glossary

Denominator
The number below the line in a fraction. This quantity is the number of units the whole group will be divided by. For example, in 6/7, 7 is the denominator.

Divisor
The quantity that is used to divide another quantity.

Elevation View
A technical drawing that has the perspective of looking from the side at the object drawn. For example, looking at a house from the street.

English System
A system of weights and measures based on the foot as the unit of length and the pound as the unit of weight,

Equation
A group of numbers and/or symbols separated into left and right sides by an equals symbol.

Estimate
To calculate approximately the amount of something.

Even Number
Any number that when divided by two has no remainder.

Fraction
The portion of a larger quantity, usually written as a numerator being divided by a denominator. For example, 1/3 is a fraction.

Geometry
Mathematical representation of measurement, and the relationship of points, lines, angles, and surfaces.

Mathematician's Glossary

Graph
A picture used to display relationships between numbers or information.

Measurement
The process of determining the quantity of something. Length, area, volume, time, and mass are the basic quantities determined.

Metric System
A standard of weights and measures based on the meter as the unit of length and the kilogram as the unit of mass.

Minus Sign
The symbol (-) that is used to indicate a negative number or subtraction.

Numerator
The number above the line in a fraction. This number represents the whole group that will be divided into equal quantities. For example, in the fraction 5/9, 5 is the numerator.

Obtuse Angle
An angle more than 90 degrees.

Odd Number
An integer that is not evenly divisible by two.

Ordinal Number
A number that indicates position in a series or order. For example, "sixth" is an ordinal number.

Mathematician's Glossary

Perimeter
A closed curve or line that surrounds an area.

Plan View
A technical drawing that has the perspective of looking down from above on the object drawn. For example, looking straight down at a house from above the house.

Probability
The chance or odds that a certain event will happen.

Ratio
Two quantities expressed as one divided by the other. For example, in a class with 10 boys and 15 girls, the ratio of boys to girls is 10/15.

Reflex Angle
An angle more than 180 degrees but less than 360 degrees.

Remainder
The number that is left over after performing division.

Right Angle
An angle equal to 90 degrees.

Right Triangle
A triangle that has one 90 degree angle.

Sum
The amount obtained as a result of adding. For example, 16 is the sum of 6 + 10.

Triangle
A three-sided closed polygon with three angles. The sum of these angles is 180 degrees.

CONVERSION TABLES

Multiply	by	To Obtain
feet	30.48	centimeters
feet	0.3048	meters
feet	0.0001894	miles
feet/min	0.5080	centimeters/sec
feet/sec	0.6818	miles/hour
gallons	0.1337	cubic feet
gallons	3.785	liters
hours	0.04167	days
hours	0.0059520	weeks
inches	2.540	centimeters
inches	0.0000158	miles
kilograms	2.205	pounds
kilometers	3281	feet
kilometers	1000	meters
kilometers	0.6214	miles
liters	61.02	cubic inches
liters	0.2642	gallons
liters	2.113	pints

Conversion Tables

Multiply	by	To Obtain
meters	100	centimeters
meters	3.281	feet
meters	0.001	kilometers
meters	0.0006214	miles
meters	1000	millimeters
miles	5280	feet
miles	1.609	kilometers
miles/hour	88	feet/minute
ounces	28.3495	grams
ounces	0.0625	pounds
pints	0.125	gallons
pints	0.5	quarts
pounds	0.4536	kilograms
pounds	16	ounces
quarts	0.25	gallons
quarts	0.9463	liters
yards	0.9114	meters
yards	0.0005682	miles

GENERAL FORMULAS

Circle

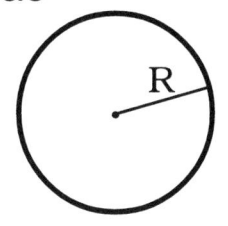

Area = 3.14 x R x R
Perimeter = 2 x 3.14 x R

Square

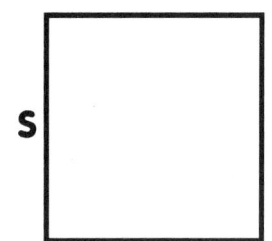

Area = S x S
Perimeter = 4 x S

Triangle

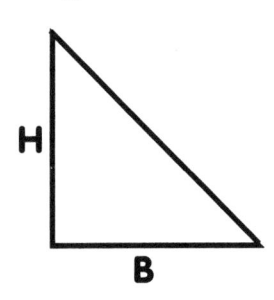

Area = 1/2 x B x H

Ellipse

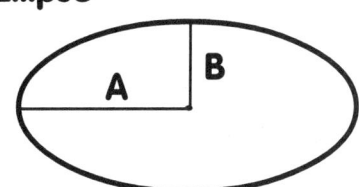

Area = 3.14 x A x B

Rectangle

Area = L x W
Perimeter = (2x L) + (2 x W)

B = Base
H = Height
L = Length
R = radius
S = side
W = Width

General Formulas

Commutative Law for Addition
A + B = B + A
This means that when adding two numbers, the order of the addition does not matter. For example, 3 + 8 = 8 + 3.

Commutative Law for Multiplication
AB = BA
This means that when multiplying two numbers, the order of multiplication does not matter. For example, 3 x 6 = 6 x 3.

Associative Law for Multiplication
A(BC) = (AB)C
This means that when multiplying three numbers, the order of the multiplication does not matter.

Distributive Law
A(B + C) = AB + AC
This means that when you multiply a sum of two numbers by a third number, it is the same as multiplying the two numbers by the third number and adding these results together.

Temperature Conversion
Celsius to Fahrenheit: (9/5 x degrees C) + 32 = Fahrenheit

Fahrenheit to Celsius: 5/9 x (degrees F - 32) = Celsius

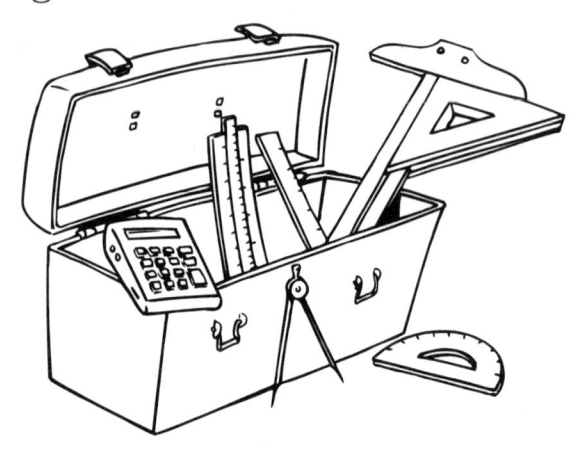

GRAPH PATTERN

PIE CHART PATTERN

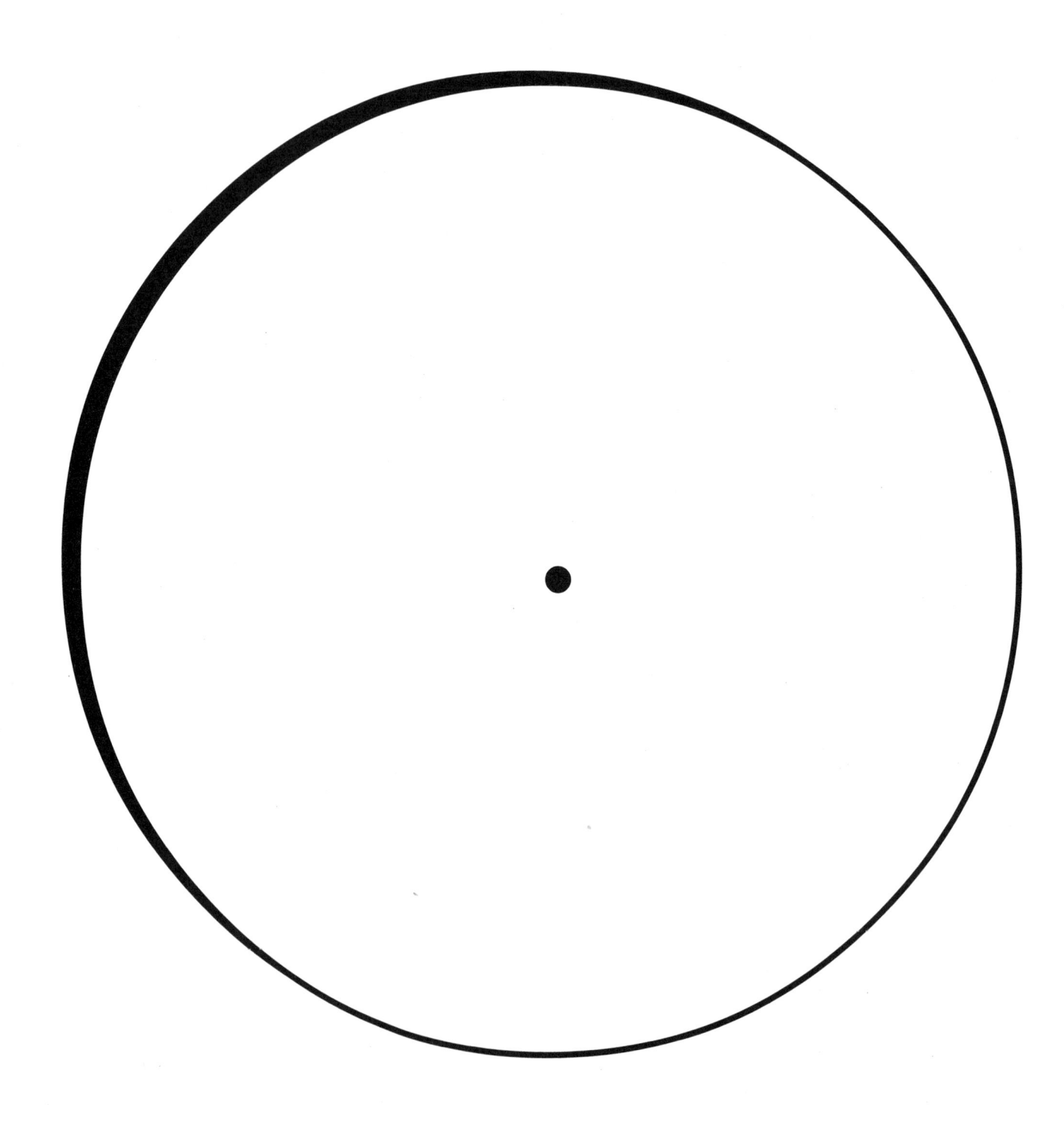

WEB EXTENSIONS

The World Wide Web is an excellent resource for many subject areas, including mathematics. Almost everything can be found on the Web.

The Web addresses (URLs) listed below can be used as starting points for research or to gather material for reports.

Center for Mars Exploration
http://cmex-www.arc.nasa.gov/
Up-to-date information on the Pathfinder and Surveyor Mars missions.

CNN Stock Quotes
http://cnnfn.com/markets/us_markets.html
A listing of stock quotes for the Dow Jones, NYSE, NASDAQ, AMEX, and S & P.

European Union Homepage
http://s700.uminho.pt/ec.html
An interactive map of Europe with flags of EU countries hot-linked to country Web Sites.

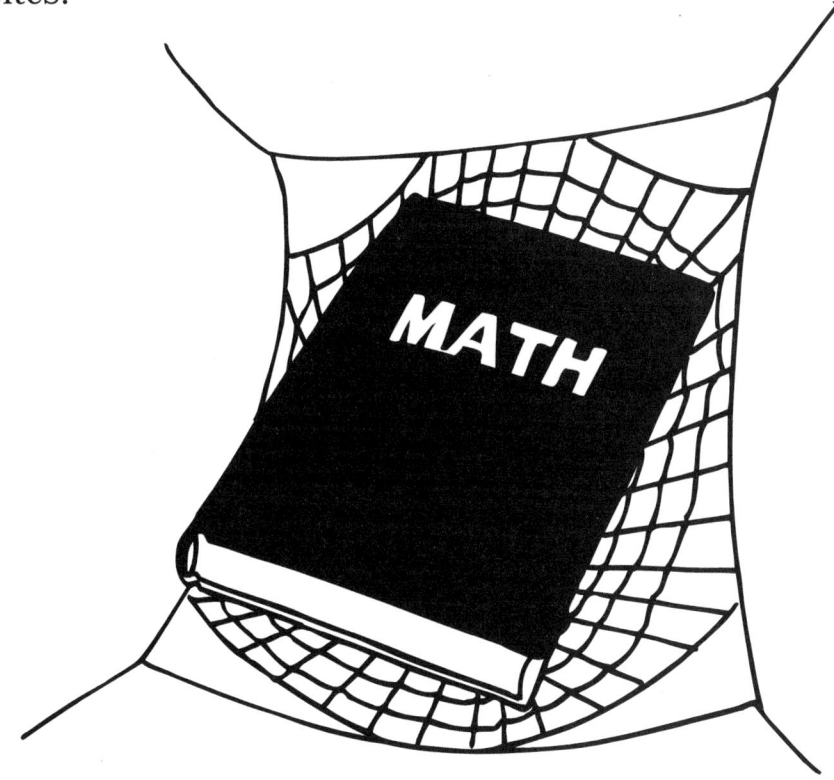

Web Extensions

Foreign Currency Exchange
http://quote.yahoo.com/forex?update
This is Yahoo's currency exchange site.

Global Positioning System Site
http://www.utexas.edu/depts/grg/gcraft/notes/gps/gps.html
Selected as the best Web Site for GPS.

Greenwich Mean Time Web Site
http://www.greenwich2000.com/time.htm
This site displays current GMT and GMT for major cities.

Haiku Archive
http://www.cs.wisc.edu/~ikunen/haiku/
A place to write, read, and learn about haiku.

Internal Revenue Service Site
http://www.irs.ustreas.gov/prod/cover.html
This IRS site has forms, tax calendars, comments, and help.

Mathematics Education Web Site
http://www.ncrel.org/msc/mathweb.htm
This page of links has lesson plans, forums, articles on improving education, and more.

US Census Bureau
http://www.census.gov/apsd/www/cqc.html
This government site has statistical information and an actual Census questionnaire.

MATH SKILLS INDEX

Many of the activities in this book incorporate several math areas, such as addition, subtraction, multiplication, and division. Additionally, graphing, percentages, and fractions are found in several activities.

Addition/Subtraction
Bony Bar Graphing 33
Buying on Credit 26
Census Report 70
Checking Account Math 8
Counting Haiku 52
Greenwich Mean Time 68
Growth Charting 31
Heartbeat Multiplication 34
Heart Rate Worksheet 35
How Far? 54
It's A Watch ... It's A Compass 67
Map Reading Math 53
Measuring with Paper 60
Savings Account Math 21
Shopping with Coupons 25
Sports Statistics 78
Syllable Counting 64
Temperature Tracking 71
Time Budget 44
Tipping Etiquette 16

Multiplication
Buying on Credit 26
Breath Count 30
Bony Bar Graphing 33
Census Report 70
Cooking with Math 19
Equator Equivalents 28
Equator Math 29
Estimating Distance 57
Eye Blinks Estimating 39
Hair Count Estimating 38

Multiplication
Heartbeat Multiplication 34
Heart Rate Worksheet 35
How Far? 54
How Many Books? 51
Income Tax Returns 23
In the Budget 6
Kid Count Estimating 50
Map Reading Math 53
Mars Mission Math 72
Metric Conversions 76
Personal Percents 36
Recipe Revision 18
Recycling Pays 12
Restaurant Math 17
Savings Account Math 21
School Days 43
Sports Statistics 78
Stock Market Math 14
Sun Protection Factoring 42
Taxing Multiplication 22
Tipping Etiquette 16
Walking Speed Calculating 27
Weather Map Data 69
25 Cents A Day 20

Math Skills Index

Division
Bony Bar Graphing 33
Computer Memory Math 77
Family Percents Graphing 47
Leg-Up Geometry 41
Measuring with Paper 60
Pace Counting Conversion 40
Personal Percents 36
Personal Percents Pie Charting 37
Sun Protection Factors 42
Walking Speed Calculating 27

Percents
Bony Bar Graphing 33
Buying on Credit 26
Currency Exchange 81
Euro Dollars 80
Family Percents Graphing 47
Income Tax Returns 23
In the Budget 6
Paying for Television 82
Personal Percents 36
Recycling Pays 12
Restaurant Math 17
Tipping Etiquette 16

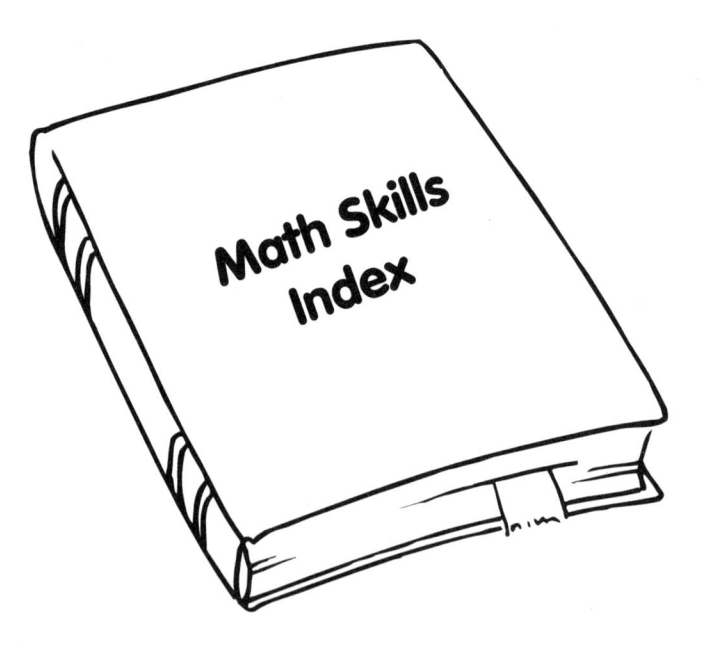

Graphing
Bony Bar Graphing 33
Family Percents Graphing 47
Favorite Foods Graphing 32
Personal Percents Pie Charting 37
Recycling Reasoning 13